Classics in Mathematics

Tonny A. Springer   Jordan Algebras and Algebraic Groups

Springer
*Berlin
Heidelberg
New York
Barcelona
Budapest
Hong Kong
London
Milan
Paris
Santa Clara
Singapore
Tokyo*

Tonny A. Springer

Jordan Algebras and Algebraic Groups

Reprint of the 1973 Edition

Springer

Tonny A. Springer
University of Utrecht
Mathematics Department
P.O. Box 80 010
3508 Utrecht
The Netherlands

QA
252.5
.S67
1998

Originally published as Vol. 75 of the
Ergebnisse der Mathematik und ihrer Grenzgebiete

Mathematics Subject Classification (1991): 17C99, 20G05

CIP data applied for

Die Deutsche Bibliothek – CIP-Einheitsaufnahme
Springer, Tonny A.:
Jordan algebras and algebraic groups / Tonny A. Springer.- Reprint of the 1973 ed.- Berlin; Heidelberg;
New York; Barcelona; Budapest; Hong Kong; London; Milan; Paris; Santa Clara; Singapore; Tokyo:
Springer, 1998
(Classics in mathematics)
ISBN 3-540-63632-3

ISSN 1431-0821
ISBN 3-540-63632-3 Springer-Verlag Berlin Heidelberg New York

This work is subject to copyright. All rights are reserved, whether the whole or part of the material is concerned, specifically the rights of translation, reprinting, reuse of illustrations, recitation, broadcasting, reproduction on microfilm or in any other way, and storage in data banks. Duplication of this publication or parts thereof is permitted only under the provisions of the German Copyright Law of September 9, 1965, in its current version, and permission for use must always be obtained from Springer-Verlag. Violations are liable for prosecution under the German Copyright Law.

© Springer-Verlag Berlin Heidelberg 1998
Printed in Germany

The use of general descriptive names, registered names, trademarks etc. in this publication does not imply, even in the absence of a specific statement, that such names are exempt from the relevant protective laws and regulations and therefore free for general use.

SPIN 10651007   41/3143-5 4 3 2 1 0 – Printed on acid-free paper

T. A. Springer

Jordan Algebras
and Algebraic Groups

Springer-Verlag Berlin Heidelberg New York
1973

Tonny Albert Springer
Mathematisch Instituut der Rijksuniversiteit te Utrecht,
De Uithof, Utrecht/Netherlands

AMS Subject Classifications (1970): 17 C 99, 20 G 05

ISBN 3-540-06104-5 Springer-Verlag Berlin Heidelberg New York
ISBN 0-387-06104-5 Springer-Verlag New York Heidelberg Berlin

This work is subject to copyright. All rights are reserved, whether the whole or part of the material is concerned, specifically those of translation, reprinting, re-use of illustrations, broadcasting, reproduction by photocopying machine or similar means, and storage in data banks. Under § 54 of the German Copyright Law where copies are made for other than private use, a fee is payable to the publisher, the amount of the fee to be determined by agreement with the publisher. © by Springer-Verlag Berlin Heidelberg 1973. Library of Congress Catalog Card Number 72-96718. Printed in Germany. Typesetting, printing and binding: Universitätsdruckerei H. Stürtz AG, Würzburg.

Preface

The aim of this book is to give an exposition of a part of the theory of Jordan algebras, using linear algebraic groups. That such groups play a role in Jordan algebra theory is a well-established fact, pointed out, for example, by H. Braun and M. Koecher in their book "Jordan Algebren". We have tried to exploit that fact as much as possible. In particular, the classification of simple Jordan algebras is derived here from the Cartan-Chevalley theory of semi-simple linear algebraic groups and their irreducible representations.

The first part of the book (until §11) is, in the main, of an elementary character. It contains part of the basic theory of finite dimensional Jordan algebras with identity. But these appear in disguise: instead of Jordan algebras we use the "J-structures", introduced in §1. The notion of a J-structure contains an axiomatization of the notion of inverse. The algebraic group which is central in the theory, the so-called structure group (introduced by Koecher), enters already in the definition of J-structures.

If the characteristic is not 2, a J-structure is essentially the same thing as a Jordan algebra (as is established in §6). One of the advantages of J-structures is that the characteristic 2 case needs no special care, at least in the elementary theory. This is not so in Jordan algebra theory, where in characteristic 2 the so-called quadratic Jordan algebras come in. The relations between these and J-structures are discussed in §7. Examples of J-structures are discussed in §2 and §5. The "quadratic map", which is familiar from Jordan algebra theory, is introduced in §3. It plays an important role. In particular, we use it in §10 for a group-theoretical version of the Peirce decomposition with respect to an idempotent element.

In §11 a classification problem for semisimple groups is solved. This solution leads quickly to the classification of simple J-structures over algebraically closed fields of characteristic not 2 in §12. The more troublesome characteristic 2 case is dealt with in §13. In §14 we then discuss the explicit determination of the structure group of the various simple J-structures, as well as the related Lie algebras (which are introduced in §4). §15 deals with the classification of J-structures over fields which are not algebraically closed. §0 contains some preliminary material, for example about polynomial and rational maps of a vector space.

The notes at the end of the sections contain various remarks and references to the literature. In the latter we have not attempted to achieve completeness. Nor does the bibliography at the end of the book claim to be comprehensive (more complete references are given in the books of Braun-Koecher and Jacobson, quoted in the bibliography).

References to the bibliography are given in square brackets. Formulas are numbered consecutively in the sections, §x, (y) means formula (y) of §x.

I am grateful to F. D. Veldkamp for a number of critical remarks and to Miss J. van der Mars for the preparation of the manuscript.

Utrecht, December 1972

T. A. Springer

Table of Contents

§ 0. Preliminaries 1
§ 1. J-structures 9
§ 2. Examples 23
§ 3. The Quadratic Map of a J-structure 39
§ 4. The Lie Algebras Associated with a J-structure 48
§ 5. J-structures of Low Degree 54
§ 6. Relation with Jordan Algebras (Characteristic $\neq 2$) 66
§ 7. Relation with Quadratic Jordan Algebras 72
§ 8. The Minimum Polynomial of an Element 79
§ 9. Ideals, the Radical 83
§ 10. Peirce Decomposition Defined by an Idempotent Element . . 90
§ 11. Classification of Certain Algebraic Groups 106
§ 12. Strongly Simple J-structures 122
§ 13. Simple J-structures 128
§ 14. The Structure Group of a Simple J-structure and the Related Lie Algebras 136
§ 15. Rationality Questions 158

Bibliography 167

Index 169

§0. Preliminaries

0.1 Algebraic geometry, algebraic groups. We shall make use of language and results of algebraic geometry and the theory of linear algebraic groups. In the first ten sections what we need is mainly elementary. In fact, the less elementary parts (e.g. the end of §2 from 2.21 on, and 4.10) are there as examples or are only needed in the later sections. From §11 on we have to use, however, the theory of semisimple groups and their root systems. The reader is then assumed to be familiar with these matters.

As a standard reference for the theory of linear algebraic groups we have used Borel's book [1]. This also contains a good résumé of algebraic geometry. [1] does not discuss the details of the theory of semisimple groups. Appropriate references will be given in the later sections.

We shall now discuss a number of elementary notions and facts, which will be needed.

0.2 Fields, vector spaces. In the sequel, K will always denote an algebraically closed field, its characteristic $\operatorname{char}(K)$ is usually denoted by p. If k is a subfield of K then k_s (resp. $\bar{k}$) denotes a separable (resp. algebraic) closure of k in K. K^* denotes the multiplicative group of K.

Let V be a (non necessarily finite dimensional) vector space over K. We denote by $GL(V)$ the group of all nonsingular linear transformations of V, this is a linear algebraic group, see [1, p. 93–94]. We denote by $\operatorname{End}(V)$ the K-algebra of all K-linear endomorphisms of V. $\operatorname{End}(V)$, equipped with its natural Lie algebra structure, is written $\mathfrak{gl}(V)$. This Lie algebra is identified with the Lie algebra of $GL(V)$, see [1, 3.6, p. 130]. $\mathbb{GL}_n$ denotes the general linear group (see [loc. cit., p. 91]). If V is an n-dimensional vector space then $GL(V)$ is isomorphic to $\mathbb{GL}_n$. $SL(V) \subset GL(V)$ is the special linear group, consisting of all linear transformations of V with determinant 1. $\mathbb{SL}_n$ is the corresponding subgroup of GL_n.

k denoting a subfield of K, a k-structure on V is a k-module $V_k \subset V$ such that the homomorphism $K \otimes V_k \to V$, induced by the inclusion, is an isomorphism. See [1, 11.1, p. 40] for further details. Occasionally, we shall use k-structures on algebraic varieties. For this notion see [1, p. 44–47].

0.3 Polynomial functions and maps. Let V be a finite dimensional vector space over K. We denote by $K[V]$ the symmetric algebra on the dual V^* of V. Recall that $K[V]$ can be defined as the quotient of the tensor algebra $T(V^*)$ on V^* by the twosided ideal generated by the elements $x \otimes y - y \otimes x$ ($x, y \in V^*$). $K[V]$ is commutative.

Let $(e_i)_{1 \leq i \leq n}$ be a basis of V, let $(x_i)_{1 \leq i \leq n}$ be the dual basis of V^*. This means that $x_i(e_j) = \delta_{ij}$ (Kronecker delta). The x_i are identified with their canonical images in $K[V]$, we then have $K[V] = K[x_1, \ldots, x_n]$. The n elements x_i are algebraically independent over K, hence $x_i \mapsto X_i$ defines an isomorphism of $K[V]$ onto the polynomial algebra $K[X_1, \ldots, X_n]$ in n indeterminates $(X_i)_{1 \leq i \leq n}$. Let

$$f = \sum_{i_1, \ldots, i_n \geq 0} a_{i_1 \ldots i_n} x_1^{i_1} \ldots x_n^{i_n}$$

be an element of $K[V]$. We identify f with the function on V defined as follows: if $x = \sum_{i=1}^{n} a_i e_i \in V$, then

(1) $$f(x) = \sum_{i_1, \ldots, i_n \geq 0} a_{i_1 \ldots i_n} a_1^{i_1} \ldots a_n^{i_n}.$$

Since $f(x) = 0$ for all $x \in V$ if and only if $f = 0$, we may and shall identify $K[V]$ with an algebra of functions on V (viz. the functions given by an expression (1)). We call the elements of $K[V]$ *polynomial functions* on V.

$f \in K[V]$ is homogeneous of degree n if $f(ax) = a^n f(x)$ for $a \in K$, $x \in V$. Such f form a subspace $K[V]^n$ of $K[V]$ and

$$K[V] = \bigoplus_{n \geq 0} K[V]^n$$

is a grading of $K[V]$.

If V has a k-structure (in the sense of 0.2) then $f \in K[V]$ is said to be *defined over k* if the following holds: taking (e_i) to be a basis of $V_k \subset V$ over k, then the coefficients $a_{i_1 \ldots i_n}$ in (1) lie in k (this is independent of the choice of the basis of V_k). Let $K[V]_k$ be the set of $f \in K[V]$ which are defined over k, then $K[V]_k$ defines a k-structure on $K[V]$.

Let W be a second finite dimensional vector space. A map $\phi: V \to W$ is called a *polynomial map* if, with respect to some basis of W, the coordinates of $\phi(x)$ are polynomial functions of $x \in V$. The polynomial maps form a vector space $K[V, W]$, which is a free $K[V]$-module, isomorphic to $K[V] \otimes_K W$. We have $K[V, K] = K[V]$. $K[V, W]$ is graded in the obvious manner. If V and W have k-structures, then there is an obvious way of defining a k-structure $K[V, W]_k$ on $K[V, W]$.

0.4 The Zariski topology. Let V be again a finite dimensional vector space over K. The Zariski topology on V is the topology whose closed sets are the algebraic subsets of V, i.e. the sets $S \subset V$ such that there exist a set $f_1, \ldots, f_d$ of polynomial functions on V, with

$$S = \{x \in V \mid f_1(x) = \cdots = f_d(x) = 0\}.$$

Topological notions will always be relative to the Zariski topology. Recall that nonempty open subsets of V are dense. Moreover any two such subsets have a nonempty intersection.

0.5 Rational functions and maps. V being as before, let $K(V)$ be the quotient field of $K[V]$. We call the elements of $K(V)$ *rational functions* on V.

Let $f \in K(V)$. There exist $g, h \in K[V]$ such that $h \neq 0$ and that

(2) $$f = g h^{-1}.$$

$K[V]$ being isomorphic to a polynomial algebra $K[X_1, \ldots, X_n]$, we have unique factorization in $K[V]$. It follows that there exists an expression (2) such that g and h have no common factor of strictly positive degree. We call this a *reduced expression* of f. g is then called a *numerator* of f and h a *denominator*. They are unique up to a nonzero scalar factor. A denominator of f is a polynomial function h of minimal degree such that (2) holds.

Let U be a nonempty open subset of V, let $K[U]$ be the ring of functions f on U such that there exist $g, h \in K[V]$ with $h(x) \neq 0$ for all $x \in U$ and
$$f(x) = g(x) h(x)^{-1} \qquad (x \in U).$$

The $K[U]$ form an inductive system (via inclusion of U) and we can then identify $K(V)$ with the inductive lim ind $K[U]$, see [1, §8, p. 35]. Hence we view rational functions on V as veritable functions defined in open subsets of V. We say that $f \in K(V)$ is *defined* in x or *regular* in x if there is an expression (2) with $h(x) \neq 0$. We then put $f(x) = g(x) h(x)^{-1}$.

Let W be a second finite dimensional vector space. We put

$$K(V, W) = K(V) \otimes_{K[V]} K[V, W],$$

we call the elements of $K(V, W)$ *rational maps* of V into W. $K(V, W)$ is a finite dimensional vector space over $K(V)$, isomorphic to $K(V) \otimes_K W$. If $\phi \in K(V, W)$ then there exists a polynomial map $\psi \in K[V, W]$ and $h \in K[V]$ such that
$$\phi = h^{-1} \psi.$$

An h of minimal degree is called a *denominator* of ϕ, it is unique up to a nonzero scalar factor. ψ is called a *numerator* of ϕ. ϕ is said to be *defined* in x or *regular* in x if $h(x) \neq 0$. We then write $\phi(x) = h(x)^{-1} \psi(x)$.

If Z is a third vector space, there is a composition map $(\phi, \psi) \mapsto \phi \circ \psi$ of $K(W, Z) \times K(V, W)$ to $K(V, Z)$, defined in an obvious manner. In particular, one can compose rational maps of V into V. $\phi \in K(V, V)$ is said to be *birational* if there exists $\psi \in K(V, V)$ such that $\psi \circ \phi = \phi \circ \psi = \mathrm{id}$. The birational maps of V form a group.

If V has a k-structure then $f \in K(V)$ is said to be *defined over k* if there is an expression (2), with g and h defined over k (caution: the $f \in K(V)$ which are defined over k do not provide a k-structure on the vector space $K(V)$). A similar definition can be given for $f \in K(V, W)$.

We next discuss some special results about rational functions and maps.

0.6 A lemma on polynomial functions. Let V be a finite dimensional vector space. Fix $a \in V$. If $f \in K[V]$ we define for each $x \in V$ a polynomial function f_x on K by
$$f_x(t) = f(t a + x).$$

Let $f_1, \ldots, f_s$ be s homogeneous polynomial functions on V, let g be a greatest common divisor of these polynomials. Since a divisor of a homogeneous polynomial function is again homogeneous, it follows that g is homogeneous.

0.7 Lemma. *There exists a nonempty open subset U of V such that for $x \in U$ we have that g_x is a greatest common divisor of the polynomial functions $(f_1)_x, \ldots, (f_s)_x$.*

An induction shows that it suffices to consider the case $s = 2$. We then may also assume $g = 1$. Let h_x be a g.c.d. of $(f_1)_x$ and $(f_2)_x$. Using Euclid's algorithm one sees that there exists a rational function h on $K \times V$ whose denominator is a polynomial function on V and a nonempty open subset $U_1 \subset V$ such that we may take
$$h_x(t) = h(t, x),$$
if $x \in U_1$. Then there exists $g_i \in K(K, V)$ $(i = 1, 2)$ such that
$$f_i(t a + x) = h(t, x) g_i(t, x),$$
if $t \in K$, $x \in U_1$. Using Gauss' lemma it follows, replacing U_1 by a suitable open subset, that we may assume that h and g_i are polynomial functions. Putting $t = 0$ we conclude that $h(0, x)$ must be a nonzero constant. Since h is homogeneous, it follows that h is constant. This implies that g_x is constant for $x \in U$, which proves 0.7.

Now let W be another finite dimensional vector space, let ϕ be a rational map of V into W. Let g be a denominator of ϕ. Fix $a \in V$. There is a nonempty open subset U_1 of V such that for $x \in U_1$ we have that
$$\phi_x(t) = \phi(t a + x),$$

§0. Preliminaries

defines a rational map of K into W. The next lemma is an immediate consequence of 0.7.

0.8 Lemma. *There is a nonempty open subset U of V such that for $x \in U$ we have that ϕ_x is a rational map of K into W with denominator g_x.*

0.9 Differentiation of rational maps. Let $K[[V]]$ be the ring of formal power series over K, let $K((X))$ be its quotient field. Since $K[X] \subset K[[X]]$ there is a canonical injection α of $K(X)$ into $K((X))$.

If W is a vector space, let $W[[X]] = W \otimes_K K[[X]]$, $W((X)) = W \otimes_K K((X))$. We write the elements of $W[[X]]$ as formal power series with coefficients in W and those of $W((X))$ as formal Laurent series, with finitely many negative powers of X, and with coefficients in W.

Now let V and W be two finite dimensional vector spaces, let $\phi \in K(V, W)$ be a rational map of V into W. There is an open subset $U \neq \emptyset$ of V such that for $x, y \in U$ we have that

$$t \mapsto \phi(x + t y)$$

defines a rational map $\phi_{x,y}$ of K into W. Since the field of rational functions on K is isomorphic to $K(X)$ (the isomorphism sending the identity map onto X), it follows that there is an isomorphism δ of $K(K, W)$ into $W((X))$. We write

$$\delta \phi_{x,y} = \phi(x + y X)$$

and we then have a formal series

$$\phi(x + y X) = \sum_{i > -s} (d_i \phi)_x(y) X^i.$$

If ϕ is a polynomial map of degree n, then

$$\phi(x + t y) = \sum_{i=0}^{n} (d^i \phi)_x(y) t^i,$$

if $x, y \in V$, $t \in K$.

Moreover we have now $(d^0 \phi)_x(y) = \phi(x)$ and $y \mapsto (d^i \phi)_x(y)$ is for fixed x a polynomial map $V \to W$, which is homogeneous of degree i. $x \mapsto (d^i \phi)_x(y)$ is for fixed y a polynomial map of degree $\leq n - i$.

Now let ϕ be an arbitrary rational map and assume that ϕ is defined in x. Let g (resp. h) denote a numerator (resp. a denominator) of ϕ. We then have the identity of formal series

$$h(x + y X) \phi(x + y X) = g(x + y X),$$

from which one finds that

(3) $$\sum_{a+b=i} (d^a h)_x(y) \cdot (d^b \phi)_x(y) = (d^i g)_x(y).$$

Since $h(x) \neq 0$, it follows that $(d^i \phi)_x(y) = 0$ if $i < 0$, moreover $(d^0 \phi)_x(y) = \phi(x)$. (3) now readily implies that $y \mapsto (d^i \phi)_x(y)$ is a polynomial map which is homogeneous of degree i.

We write $(d\phi)_x(y) = (d^1 \phi)_x(y)$ and we call this the *derivative of ϕ in x, in the direction y*. $(d\phi)_x$ is a linear map $V \to W$, the *differential* of ϕ in x. $x \mapsto (d\phi)_x$ is a rational map of V into the space of endomorphisms of V. The $(d^i \phi)_x(y)$ (resp. the rational maps $x \mapsto (d^i \phi)_x$) with $i \geq 2$ are called the *higher differentials* of ϕ in x, in the direction y (resp. the higher differentials of ϕ).

The proof of the next two lemma's is left to the reader.

0.10 Lemma. *Let f be a rational function on V, let ϕ be a rational map of V into W. Assume that f and ϕ are defined in x. Then $f\phi$ is defined in x and*

$$d(f\phi)_x(y) = f(x) \cdot (d\phi)_x(y) + (df)_x(y) \cdot \phi(x).$$

It follows from 0.10 that, if $y \in V$ is fixed, the endomorphism D_y of $K[V]$ given by
$$(D_y f)(x) = (df)_x(y)$$
is a derivation of $K[V]$.

0.11 Lemma *(chain rule). Let V, W, Z be three finite dimensional vector spaces. Let ϕ and ψ be rational maps of W into Z and of V into W, respectively. Suppose that ψ is defined in $x \in V$ and that ϕ is defined in $\psi(x) \in W$. Then the composite $\phi \circ \psi$ is defined in x and*

$$\left(d(\phi \circ \psi)\right)_x = (d\phi)_{\psi(x)} \circ (d\psi)_x.$$

0.12 Lemma. *Let f be a rational function on V such that $(df)_x = 0$ whenever ϕ is defined in x. Then there exists a rational function f_1 on V such that $f = f_1^p$, where $p = \mathrm{char}(K)$. In particular, f is a constant polynomial map if $p = 0$.*

Let g (resp. h) denote a numerator (resp. a denominator) of f. Since we have the usual formula for the derivative of quotients

$$\left(d(h^{-1}g)\right)_x(y) = h(x)^{-1}(dg)_x(y) - h(x)^{-2}(dh)_x(y) \cdot g(x),$$

as follows from 0.10, the assumption implies that

$$h(x) \cdot (dg)_x(y) = (dh)_x(y) \cdot g(x),$$

whenever $h(x) \neq 0$, hence for all $x, y \in V$. This implies that the polynomial function $x \mapsto (dh)_x(y)$, whose degree is less then that of h, must be divisible by h. Consequently $(dh)_x(y) = 0$. It is well-known that then h is a p-th power (for a related result see [5, Prop. 4, p. 73]). Similarly g is a p-th power, whence the assertion.

0.13 Lemma. *Let f be a polynomial function on V with the following property: there is a nonempty open set $U \subset V$ such that for all $x \in U$ and*

§0. Preliminaries

all $y \in V$ the polynomial function

$$t \mapsto f(tx+y)$$

is a power of a linear function of t. Then f is a power of a polynomial function of degree 1.

We may assume f to be irreducible. Fix $x \in U$, $y \in V$. Let

$$f(tx+y) = (at+b)^n,$$

where $a \in K^*$, $b \in K$. Then

$$a^n = f(x), \quad n\,a^{n-1} b = (df)_x(y).$$

If n were divisible by the characteristic p of K, we had $(df)_x(y) = 0$ for all $x, y \in V$. Then f would have to be a p-th power. This a contradiction. Hence we have

$$a^{-1} b = n^{-1} f(x)^{-1} (df)_x(y).$$

It follows that

$$f(tx+y) = f(x)\bigl(t + n^{-1} f(x)^{-1} (df)_x(y)\bigr)^n.$$

Putting $t = 0$ we obtain the assertion, since $(df)_x(y)$ depends linearly on y.

A rational map $\phi \colon V \mapsto W$ is said to be *homogeneous of degree h* if $\phi(ax) = a^h \phi(x)$, if $a \in K^*$, $x \in V$.

0.14 Lemma (*Euler's differential equation*). *If ϕ is a rational map of V into W which is homogeneous of degree h then*

$$(d\phi)_x(x) = h\phi(x),$$

if ϕ is defined in x.

We have $\phi((t+1)x) = (t+1)^h \phi(x)$ if $t \neq -1$ and if ϕ is defined in x. Hence, X denoting an indeterminate,

$$\phi(x + xX) = (X+1)^h \phi(x),$$

which implies the assertion.

0.15 Degeneracy of polynomial functions. $f \in K[V]$ is said to be a *degenerate* polynomial function if there exists $a \neq 0$ in V such that

(4) $$f(x+a) = f(x) \quad (x \in V).$$

Otherwise f is said to be *nondegenerate*. Degeneracy of f means that there exists $a \neq 0$ such that we have for all $i \geq 1$

$$(d^i f)_x(a) = 0 \quad (x \in V).$$

In particular, we then have $(df)_x(a) = 0$ for all $x \in V$. Conversely, if this is so and if $\operatorname{char}(K) = 0$, it is easily seen that (4) holds.

0.16 Quadratic forms. A polynomial function Q on V which is homogeneous of degree 2 is called a *quadratic form*. Then
$$Q(x, y) = Q(x+y) - Q(x) - Q(y)$$
defines a symmetric bilinear form $Q(\ ,\ )$ on $V \times V$, called the *associated bilinear form*. We have
$$Q(x) = 2Q(x, x).$$

If $\operatorname{char}(K) \neq 2$, then Q is degenerate if and only if the associated bilinear form is degenerate, i.e. if there exists $a \neq 0$ such that $Q(x, a) = 0$ for all $x \in V$.

If $\operatorname{char}(K) = 2$ then Q is said to be *defective* (resp. *nondefective*) if the associated bilinear form is degenerate (resp. nondegenerate). Since we now have $Q(x, x) = 0$, the associated bilinear form is alternating. One then knows that $\dim V$ is even if the associated bilinear form is nondegenerate. If $\operatorname{char}(K) = 2$ and if Q is nondegenerate and defective, then one shows that the subspace of $a \in V$ such that
$$Q(x+a) = Q(x) \quad (x \in V)$$
has dimension 1 (using that we work over an algebraically closed field). It follows that then $\dim V$ is odd. For these elementary facts about quadratic forms in characteristic 2 see [12, p. 33].

0.17 Algebras. By an algebra A we understand a finite dimensional vector space over K, together with distributive bilinear multiplication $(a, b) \mapsto ab$, which is not necessarily associative. We always assume the existence of an identity element.

Let $k \subset K$ be a subfield, let A_k be a k-structure on the algebra A. Then $A_k \times A_k$ is a k-structure on $A \times A$. We say that the algebra A is defined over k is the polynomial map $\phi: A \times A \to A$ with $\phi(a, b) = ab$ is defined over k in the sense of 0.3 and if the identity element of A is rational over k, i.e. lies in A_k.

Let $\mathbb{M}_n(K)$ be the associative algebra of $n \times n$ matrices over K. A *central simple associative algebra over* $k \subset K$ is an algebra defined over k, which is K-isomorphic to $\mathbb{M}_n(K)$. n is then called the degree of A.

It is clear how to define the notion of a twosided ideal in an algebra A. A is called *simple*, if $\{0\}$ and A are the only twosided ideals. We shall not use a notion of k-simplicity for algebras A defined over a field $k \subset K$, i.e. for us simplicity will always mean what is called "absolute" simplicity.

Notes

The material about polynomial and rational maps and their differentiation is also discussed in [8, p. 60–64] and [14, p. 214–220]. The treatment given here, using formal power series, is somewhat different from that given in these references.

§1. J-structures

Let V be a finite dimensional vector space, let j be a rational map $V \to V$. Denote by n and N a numerator and a denominator of j, respectively. n is a polynomial map of V into V and N a polynomial function on V (see 0.5). Let H be the subset of $GL(V) \times GL(V)$ consisting of the pairs (g, h) such that

(1) $$g \circ j = j \circ h.$$

1.1 Lemma. *H is a closed subgroup of $GL(V) \times GL(V)$.*

Clearly H is a subgroup of $GL(V) \times GL(V)$. Let $(g, h) \in H$. n and N being as above, we have

(2) $$N(hx) g(nx) = N(x) n(hx),$$

for all $x \in V$. Let $(e_i)_{1 \leq i \leq m}$ be a basis of V. Put

$$g e_i = \sum_{i=1}^{m} t_{ij} e_j,$$

$$h e_i = \sum_{i=1}^{m} u_{ij} e_j,$$

and write

$$x = \sum_{i=1}^{m} x_i e_i.$$

Then (2) is equivalent to a finite family of polynomial relations between the x_h, t_{ij}, u_{rs}, which have to hold for all x_i. It follows that $(g, h) \in H$ if and only if t_{ij} and u_{rs} satisfy a finite set of polynomial relations. This proves that H is a closed subgroup of $GL(V) \times GL(V)$.

1.2. Let $\pi: GL(V) \times GL(V) \to GL(V)$ be projection on the first factor. j and H being as above, the projection πH is a closed subgroup of $GL(V)$, which is called the *structure group* of j and which will be denoted by $G(j)$. $G(j)$ consists of the $g \in GL(V)$ for which there exists $h \in GL(V)$ such that (1) holds. If j is defined over k, then 1.1 implies that $G(j)$ is k-closed in the sense of [1, p. 42].

Suppose now that j is a birational map $V \to V$. If $g \in G(j)$, there exists $h \in GL(V)$ such that (1) holds and the birationality of j shows that h is unique. It follows from (1) that now

$$h \circ j^{-1} = j^{-1} \circ g,$$

hence $h \in G(j^{-1})$. Put $h = \alpha(g)$. Clearly α is a group homomorphism of $G(j)$ to $G(j^{-1})$. In fact, α is a morphism of algebraic groups. For $\alpha(g) = j^{-1} \circ g \circ j$ implies that α is a rational map $G(j) \to G(j^{-1})$, hence a morphism of algebraic groups, as follows from [1, 1.3(a), p. 87]. Interchanging the roles of j and j^{-1}, we get a morphism $\beta: G(j^{-1}) \to G(j)$, with $\beta(h) = j \circ h \circ j^{-1}$. α and β are clearly inverses of each other, hence α is an isomorphism.

In particular, if j is involutorial, i.e. $j \circ j = \mathrm{id}$, then $G(j) = G(j^{-1})$ and α is an automorphism of the algebraic group $G(j)$.

1.3 Definition of J-structures. A *J-structure* is a triple $\mathscr{S} = (V, j, e)$, where V is a finite dimensional vector space, j a birational map of V and e a nonzero element of V, satisfying axioms to be stated presently. In these we need the structure group $G(j)$ of the birational map j which will be denoted by $G_{\mathscr{S}}$ or G. We call the linear algebraic group $G_{\mathscr{S}}$ the *structure group* of the J-structure.

The axioms are as follows:

(J1) (i) *j is a homogeneous birational map of V of degree -1 and $j = j^{-1}$,*
(ii) *j is regular in e and $je = e$.*

(J2) *If $x \in V$ is such that j is regular in x, $e + x$ and $e + jx$, then*

(3) $$j(e+x) + j(e+jx) = e.$$

(J3) *The orbit Ge of e under the structure group G is Zariski-open in V.*

It would amount to the same thing to require this orbit to be Zariski-dense, see [3, Proposition, p. 98].

We say that $\mathscr{S}$ is a *J-structure over a subfield k of K* or that $\mathscr{S}$ is *defined over k* if there exists a k-structure on the vector space V such that j is defined over k and that $e \in V(k)$. In that case G is k-closed.

Let $\mathscr{S} = (V, j, e)$ and $\mathscr{S}' = (V', j', e')$ be two J-structures. A *morphism* of $\mathscr{S}$ to $\mathscr{S}'$ is a linear map $f: V \to V'$ such that (i) $f \circ j = j' \circ f$, (ii) $f(e) = e'$. f is an *isomorphism* of J-structures if it is a bijective linear map. Automorphisms are defined in the obvious way. $\mathscr{S}'$ is a *J-substructure* of $\mathscr{S}$ if V' is a subspace of V, if $e' = e$ and if the injection $V' \to V$ is a morphism of J-structures. Then j' is the restriction of j to V'.

If $\mathscr{S}$ and $\mathscr{S}'$ are J-structures over k, then a morphism $f: \mathscr{S} \to \mathscr{S}'$ is said to be a *k-morphism* or to be *defined over k* if the linear transformation f is defined over k.

§1. J-structures

By 1.2 there is an automorphism $\alpha\colon g\mapsto g'$ of the linear algebraic group G, such that
$$g\circ j = j\circ g'.$$
We call α the *standard automorphism* of G. We have $\alpha^2 = \mathrm{id}$.

For $t\in K^*$, let
$$s_t(x) = tx \quad (x\in V)$$
be scalar multiplication by t. From the homogeneity of j ((J1)(i)) it follows that $s_t \in G$, and that
$$(s_t)' = s_{t^{-1}}.$$

A subspace I of V is called an *ideal* of $\mathscr{S}$ if for each $i\in I$ the rational map
$$x\mapsto j(x+i)-j(x),$$
which is defined on a suitable nonempty open subset of V, is a rational map of V into I. $\mathscr{S}$ is called *simple* if $\{0\}$ and V are the only ideals of $\mathscr{S}$. We shall discuss ideals in §9.

1.4 The norm. Let $\mathscr{S} = (V, j, e)$ be a J-structure. Denote again by n and N a numerator and denominator of j. n is a homogeneous morphism $V\to V$ and N a homogeneous polynomial function $V\to K$. j is regular in $x\in V$ if and only if $N(x)\neq 0$. If this is so, we say that x is *invertible* in $\mathscr{S}$ or in V.

Let d be the degree of N, then the degree of n is $d-1$ (since j is homogeneous of degree -1). We call d the *degree* of $\mathscr{S}$. By (J2)(ii), we have $N(e)\neq 0$. Hence we may and shall normalize n and N by requiring $N(e)=1$. Then n and N are uniquely determined. We call N, thus normalized, the *norm* of $\mathscr{S}$. If necessary, we shall write $N_\mathscr{S}$ (and $n_\mathscr{S}$).

1.5 Proposition. *There exists a character $a\colon G\to K^*$ of the algebraic group G such that for $g\in G$ we have*

(4) $\qquad N(g\,x) = a(g)\,N(x) \quad (x\in V).$

Let $g\in G$, let $g' = \alpha(g)$. By (1) we have

(5) $\qquad N(x)^{-1} g'(n\,x) = N(g\,x)^{-1} n(g\,x).$

$N\circ g$ is a polynomial function of degree d, which is a denominator of the rational function of x, given by the right-hand side of (5). Then by (5) N is also a denominator of that rational function, whence a relation of the form (4), with a nonzero constant $a(g)$. Putting $x = e$, we obtain $a(g) = N(g\,e)$, from which it follows that $g\mapsto a(g)$ is a morphism $G\to K^*$. That a is a homomorphism is clear.

1.6 Corollary. *If $x\in V$ is invertible and if $g\in G$, then $g\,x$ is invertible.*

This follows from (5), since j is regular at x if and only if $N(x)\neq 0$.

We now shall establish a number of fundamental properties of N, which are consequence of the axioms (J2) and (J3).

1.7 Proposition. *If $x \in V$ is invertible, then*

(6) $$N(x)N(e+jx) = N(e+x).$$

By (J2) we have on a nonempty open subset of V,

$$N(e+x)^{-1} n(e+x) + N(e+jx)^{-1} n(e+jx) = e.$$

Replacing x by $t^{-1}x$ ($t \in K^*$) this gives, using homogeneity of j, that

(7) $$tN(te+x)^{-1} n(te+x) + N(e+tjx)^{-1} n(e+tjx) = e.$$

This should be viewed as an identity of rational functions on $K \times V$. Now by 0.8, there is an open subset $U \neq \emptyset$ of V such that for $x \in U$ the rational function on K defined by

$$t \mapsto N(te+x)^{-1} n(te+x)$$

has a denominator of the same degree d as N. If $N(x) \neq 0$, the same is true for the denominator of the rational function

$$t \mapsto tN(te+x)^{-1} n(te+x).$$

But from (7) it follows that this last rational function can also be written as a quotient of two polynomial functions, with a denominator which divides the polynomial function $t \mapsto N(e+tjx)$. Since this function has degree d if $N(jx) \neq 0$, we conclude that there is a constant $c \in K^*$, independent of t, such that

$$N(te+x) = cN(e+tjx),$$

for x in a nonempty open subset of V.
In this polynomial identity in t put $t=0$. We conclude that $c = N(x)$. Putting $t=1$ we obtain (6), for x in a suitable nonempty open subset of V. By continuity (6) then holds whenever j is regular in x.

1.8 Corollary. *If x is invertible we have $N(jx) = N(x)^{-1}$.*

Replacing x by tx in (6) ($t \in K^*$) and using homogeneity of j and N we obtain the equality of polynomial functions in t

$$N(x)N(te+jx) = N(e+tx),$$

is x is invertible. Taking $t=0$ the assertion follows.

1.9 Corollary. *Let $\alpha: g \mapsto g'$ be the standard automorphism of G, let a be the character of G defined in 1.5. Then $a(g') = a(g)^{-1}$.*

§1. J-structures

By 1.5 and 1.8 we have

$$a(g')N(x) = N(g' \cdot x) = N(j(g' \cdot x))^{-1} = N(g(jx))^{-1} = a(g)^{-1}N(jx)^{-1}$$
$$= a(g)^{-1}N(x),$$

whence the assertion.

1.10 Theorem. *If $x, y \in V$ are invertible, then*

(8) $$N(x)N(jx+jy)N(y) = N(x+y).$$

Let $g \in G$, let g' be as in 1.9. By 1.5, 1.7 and 1.9 we have

$$N(ge)N(g' \cdot e + j(gy))N(gy) = a(g)N(e+jy)N(y) = a(g)N(e+y)$$
$$= N(ge+gy),$$

if y is invertible. This establishes (8) with $x = ge$ instead of y. By axiom (J 3) and by 1.6, it then follows that (8) holds for x in a nonempty open subset of V and all invertible y. This implies (8), by continuity.

1.11. Let $x \in V$ be invertible, let $y \in V$. Put

(9) $$\Phi(x, y) = N(x)N(jx+y).$$

Φ is a rational function on $V \times V$ whose denominator is independent of the second variable y. For fixed x we have that $y \mapsto \Phi(x, y)$ is a polynomial function on V of degree d.

By 1.8 and 1.10, with jy instead of y, we see that

$$\Phi(x, y) = \Phi(y, x),$$

if x and y are invertible. However it then follows that the denominator of Φ does not depend on the first variable x either, consequently Φ is a polynomial function on $V \times V$, symmetric in its two variables, and of degree d (= degree N).

We put

$$\Phi(x, y) = \sum_{i=0}^{d} \Phi_i(x, y),$$

where Φ_i is a symmetric polynomial function on $V \times V$, which is homogeneous of degree i in either of its variables. Taking $y = 0$ and using 1.8 we find that $\Phi_0 = 1$.

Let $g \in G$. By 1.5 and 1.9 we have

$$\Phi(gx, g'y) = \Phi(x, y),$$

whence

$$\Phi_i(gx, g'y) = \Phi_i(x, y).$$

We put $\Phi_1 = \sigma$ (or $\sigma_{\mathscr{S}}$, if necessary). Then σ is a symmetric bilinear form on $V \times V$, and we have for $g \in G$

(10) $$\sigma(g\,x, g'\,y) = \sigma(x, y).$$

We call σ the *standard symmetric bilinear form* of $\mathscr{S}$. The linear function τ on V with

$$\tau(x) = \sigma(x, e)$$

is called the *trace* of $\mathscr{S}$. We have

(11) $$\sigma(x, y) = N(x)(dN)_{jx}(y),$$

if x is invertible, for all y. This follows from (9) (recall the definition of $(dN)_{jx}$, see 0.9).

We next make a digression, which is suggested by the properties of N which we just discussed. We denote by $G°$ the identity component of G, i.e. the connected component of the neutral element of G.

1.12 Proposition. *Let F be an irreducible factor of the polynomial function N. Then F is a homogeneous polynomial function. There exists a character $c: G° \to K^*$ of the algebraic group $G°$ such that*

(12) $$F(g\,x) = c(g)\,F(x) \qquad (g \in G°, x \in V).$$

If $F(e) = 1$ we have

(13) $$F(x)\,F(j\,x + j\,y)\,F(y) = F(x+y),$$

(14) $$F(j\,x) = F(x)^{-1},$$

for all invertible $x, y \in V$.

Let $N = \prod_{i=1}^{h} F_i$ be a decomposition of N as a product of irreducible polynomial functions. F equals one of the F_i, up to a scalar factor. It follows from the uniqueness of the decomposition that the elements of $G°$ permute the F_i up to a nonzero scalar. $G°$ being connected, the resulting homomorphism of $G°$ into the group of permutations of the F_i must be trivial. This implies (12). It also follows that F_i is homogeneous.

x and y being invertible, let $G_i(x, y) = F_i(j(j\,x + j\,y))$ ($1 \leq i \leq h$). G_i is a rational function on $V \times V$. From the definition of N and the homogeneity of j and F_i we conclude that a numerator of G_i must divide a power of the polynomial function $(x, y) \mapsto N(x)\,N(y)$ on $V \times V$. It follows that there exist homogeneous polynomial functions Q_i and R_i on V, and a homogeneous function S_i on $V \times V$, such that

(15) $$F_i(j(j\,x + j\,y)) = Q_i(x)\,R_i(y)\,S_i(x, y)^{-1}$$

§1. J-structures

is a reduced expression (in the sense of 0.5). We may assume that $F_i(e) = Q_i(e) = R_i(e) = 1$.

Replace x by $t^{-1}x$ ($t \in K^*$) in (15). The resulting rational function of t is regular for $t = 0$ and has there the value $F_i(y)$, as one sees by looking at the left-hand side of (15). Working with the right-hand side one concludes that we must have $F_i = R_i$. Similarly $F_i = Q_i$. It then follows from (8) that

$$\prod_{i=1}^{h} S_i(x, y) = N(x+y).$$

But then there exists a homogeneous polynomial function T_i on V such that $S_i(x, y) = T_i(x+y)$. The same argument as before, viz. replacing x by $t^{-1}x$ and putting $t = 0$, then shows that $F_i = T_i$. This implies that

$$F_i(j(jx+jy)) F_i(x+y) = F_i(x) F_i(y).$$

In this formula take $y = e$. Using axiom (J2) we obtain

$$F_i(e - j(e+x)) F_i(e+x) = F_i(x),$$

which implies that

$$F_i(x) F_i(e+jx) = F_i(e+x),$$

for all invertible x. (14) with $F = F_i$ is now proved as the similar statement of 1.8. (13) then also follows.

We say that a homogeneous polynomial function F on V is a *semi-invariant* of $\mathscr{S}$ if there exists a character c of G° such that (12) holds. We next establish a useful property of semi-invariants. For this we need the following simple lemma.

1.13 Lemma. *The orbit $G^\circ \cdot e$ is Zariski open in V.*

G° is a normal subgroup of G of finite index. Let $G = \bigcup_{i=1}^{r} g_i G^\circ$. Since Ge is open in V by axiom (J3), it is dense in V. Hence one of the sets $g_i G^\circ \cdot e$ must be dense in V, and then $G^\circ \cdot e$ is dense in V. This is equivalent to $G^\circ \cdot e$ being open, by [1, Proposition, p. 98].

1.14 Proposition. *Let $p = \operatorname{char}(K)$. Let F be a semi-invariant of S which is not a p-th power of a polynomial function. Then $(dF)_e \neq 0$.*

Let F be a semi-invariant such that $(dF)_e = 0$. It follows from (12) that

$$(dF)_{ge}(gx) = c(g)(dF)_e(x) = 0 \qquad (g \in G^\circ).$$

It follows that $(dF)_x = 0$ for all $x \in G^\circ \cdot e$. 1.13 then implies that $(dF)_x = 0$ for all x, which means that F is a p-th power of a polynomial function, see 0.12. This proves 1.14.

1.15 The differential of j. Suppose $x \in V$ is invertible. The differential $(dj)_x$ of j at x is a linear transformation on V. Since $j = j^{-1}$, whence $j \circ j = \mathrm{id}$, the chain rule for differentials (see 0.11) shows that

$$(dj)_{jx} \circ (dj)_x = \mathrm{id},$$

hence $(dj)_x$ is invertible. We put

(16) $$P(x) = -(dj)_x^{-1}.$$

P is a rational map of V into $GL(V)$ (if necessary we write $P_{\mathscr{S}}$ instead of P). P is regular in x if x is invertible and then

$$P(j\,x) = P(x)^{-1}.$$

In the next proposition we collect some properties of P, which are fairly direct consequences of the axioms. $g \mapsto g'$ is the standard automorphism of G.

1.16 Proposition. *Assume that x is invertible.*
  (i) *If $g \in G$ then P is regular in $g\,x$ and $P(g\,x) = g \circ P(x) \circ (g')^{-1}$;*
  (ii) *If $t \in K^*$ then P is regular in $t\,x$ and $P(t\,x) = t^2\,P(x)$;*
  (iii) $P(x)j\,x = x$;
  (iv) $P(e) = \mathrm{id}$;
  (v) *$P(x)$ is contained in the identity component G° of the structure group G and $P(x)' = P(x)^{-1}$;*
  (vi) *If y is invertible then $P(P(y)\,x) = P(y)P(x)P(y)$.*

We have $g' \circ j = j \circ g$. By the chain rule for differentials 0.11, this implies that

$$g' \circ (dj)_x = (dj)_{gx} \circ g,$$

whence

$$g^{-1} \circ P(g\,x) = P(x) \circ (g')^{-1},$$

which proves (i). (ii) is the particular case of (i) where g is scalar multiplication s_t ($t \in K^*$). (iii) follows from $(dj)_x(x) = -j\,x$, which is a consequence of the homogeneity of j (see 0.14).

To prove (iv) we use axiom (J2) with $t\,x$ instead of x ($x \in K^*$), which gives the identity of rational functions on $K \times V$

$$j(e + t\,x) + t\,j(t\,e + j\,x) = e.$$

Let X be an indeterminate over K. For x in a suitable nonempty open subset of V we then have the identity of formal power series (see 0.9)

(17) $$j(e + x\,X) + j(e\,X + j\,x)\,X = e.$$

The series for $j(e + x\,X)$ starts off with

$$e + (dj)_e(x)\,X,$$

§1. J-structures

and that for $j(eX+jx)X$ starts off with xX. Comparing coefficients in (17) we conclude that

$$(dj)_e(x) = -x,$$

which is (iv).

By (i) and (iv) we have

$$P(g\,e) = g(g')^{-1} \in G,$$

so that $P(x) \in G$ if x is in the open subset Ge of V. Hence P is a rational map of V into G and then $P(x) \in G$ whenever P is regular in x, in particular if x is invertible. Now the set of the x where P is regular is an open subset U of V, which is irreducible hence connected. Consequently $P(U)$ is a connected subset of G, which contains the neutral element of G (by (iv)). Hence $P(U) \subset G^\circ$.

To prove the last point of (v), observe that

$$P(g\,e)' = \bigl(g(g')^{-1}\bigr)' = g' \cdot g^{-1} = P(g\,e)^{-1},$$

so that $P(x)' = P(x)^{-1}$ on a nonempty open subset of V, hence whenever P is regular in x. Finally, (vi) is a consequence of (i) and (v).

In §3 we shall return to the properties of P and establish that P can be extended to a quadratic map of V into the space $\mathrm{End}(V)$ of endomorphisms of V.

1.17. We next give some consequences of 1.16. First we recall the definition of the higher differentials of N in e. These are given by

$$N(e+x) = \sum_{i=0}^{d} (d^i N)_e(x),$$

where $(d^i N)_e$ is a polynomial function on V of degree i (see 0.9). $(d^2 N)_e$ is a quadratic form on V. We put

$$(d^2 N)_e(x, y) = (d^2 N)_e(x+y) - (d^2 N)_e(x) - (d^2 N)_e(y).$$

Then $(x, y) \mapsto (d^2 N)_e(x, y)$ is a symmetric bilinear form on $V \times V$. Recall that we write $(dN)_e = (d^1 N)_e$. We can now state the next result.

1.18 Proposition. *Let σ be the standard symmetric bilinear form of $\mathscr{S}$. Then*

$$\sigma(x, y) = (dN)_e(x)(dN)_e(y) - (d^2 N)_e(x, y).$$

We have by 1.11, if x is invertible,

(18) $\quad N(x)N(jx+y) = 1 + \sigma(x, y) + \text{terms of degree} \geq 2 \text{ in } x \text{ and } y.$

In (18) we replace x by $e+tx$ ($t \in K$). x being fixed, we then obtain a polynomial function

$$(t, y) \mapsto N(e+tx)N(j(e+tx)+y)$$

on $K \times V$. We shall prove 1.18 by determining the terms in both sides of (18) which are bilinear in (t, y). In the right-hand side this is clearly $t\,\sigma(x, y)$.

To deal with the left-hand side, we introduce an indeterminate X over K. Consider the formal power series $j(e + xX)$ (see 0.9). By 1.16 (iv) we have

$$j(e + xX) = e - xX + \text{terms involving higher powers in } X.$$

Then

$$N(e+xX)N(j(e+xX)+y) = (1+(dN)_e(x)X+\cdots)N(y+e-xX+\cdots)$$
$$= (1+(dN)_e(x)X+\cdots)(1+(dN)_e(y-xX)+(d^2N)_e(y-xX)+\cdots)$$
$$= (1+(dN)_e(x)X+\cdots)(1+(dN)_e(y)-(dN)_e(x)X-(d^2N)_e(x,y)X+\cdots)$$
$$= 1+(dN)_e(y)+((dN)_e(x)(dN)_e(y)-(d^2N)_e(x,y)X+\cdots).$$

Now by 1.11 this is a polynomial in X (x and y being fixed). Inserting t for X we see that the bilinear term in the left-hand side of (18) is

$$t\bigl((dN)_e(x)\,dN_e(y) - (d^2N)_e(x, y)\bigr),$$

which establishes 1.18, in view of what we said before.

1.19 The inner structure group. By 1.16(v) we have $P(x) \in G^\circ$ if x is invertible. Let $G_1 \in G^\circ$ be the subgroup of G° generated by all such $P(x)$. We call G_1 the *inner structure group* of $\mathscr{S}$.

1.20 Proposition. *G_1 is a closed, connected, normal subgroup of the structure group G.*

Let $U \subset V$ be the open subvariety consisting of all invertible elements. U is irreducible. By 1.16(v), P defines a morphism $U \to G^\circ$. It then follows from a well-known result (see [1, 2.2, p. 106]) that the subgroup G_1 generated by $P(U)$ is closed and connected. That G_1 is normal in G follows from the formula

$$g\,P(x)\,g^{-1} = P(g\,x)\,P(g' \cdot e) \qquad (g \in G,\ x \text{ invertible})$$

which is a consequence of 1.16(i) and 1.16(iv).

1.21 Proposition. *Let $\mathscr{S} = (V, j, e)$ and $\mathscr{S}' = (V', j', e')$ be two J-structures, let $f: \mathscr{S} \to \mathscr{S}'$ be a morphism. There is a nonempty open set U in V consisting of invertible elements such that the elements of $f(U)$ are invertible and that*

$$P_{\mathscr{S}'}(fx) \circ f = f \circ P_{\mathscr{S}}(x), \qquad x \in U.$$

Let U' be the open subset of V' where j' is regular. $f^{-1}(U')$ is open in V, let U be the set of invertible elements in $f^{-1}(U')$. U is open and non-

§1. J-structures

empty (since $e \in U$). From $f \circ j = j' \circ f$ we obtain, using the chain rule 0.11 for differentials, that

$$f \circ (dj)_x = (dj')_{fx} \circ f \quad (x \in U),$$

which implies the assertion.

1.22 Invariant bilinear forms. Let $\mathscr{S} = (V, j, e)$ be a J-structure. We use the previous notations. A bilinear form B on $V \times V$ is called *invariant* (with respect to $\mathscr{S}$) if we have

(19) $$B(g\,x, g' \cdot y) = B(x, y),$$

if $x, y \in V$, $g \in G^\circ$, where $g \mapsto g'$ denotes the standard automorphism of G (which stabilizes G°).

1.23 Lemma. *An invariant bilinear form B is symmetric.*

From (19) it follows that

$$B(g\,e, g' \cdot e) = B(e, e).$$

Since $g' \cdot e = j(g\,e)$, we conclude that

(20) $$B(x, j\,x) = B(e, e),$$

if x is in a suitable open subset U of V which contains e.

Differentiation of (20) gives

$$B(x, (dj)_x(y)) + B(y, j\,x) = 0.$$

From the definition (16) of $P(x)$ and (19) it follows that

$$B(P(x)^{-1} x, y) = B(y, j\,x).$$

Since $P(x)^{-1} x = j\,x$ by 1.16(iii) we conclude that

$$B(j\,x, y) = B(y, j\,x),$$

for x in a suitable nonempty open subset of V. This implies the assertion.

The standard symmetric bilinear form σ of 1.14 is invariant, but might be 0. We shall now show how to obtain nonzero invariant forms.

1.24 Proposition. *Let F be an irreducible factor of the norm N of $\mathscr{S}$. There exists a nonzero invariant symmetric bilinear form B on $V \times V$ such that*

(21) $$B(x, y) = F(x)(dF)_{jx}(y),$$

if x is invertible, for all $y \in V$.

Define a rational function B on $V \times V$ by

$$B(x, y) = F(x)(dF)_{jx}(y),$$

(x invertible). Clearly B is linear in its second argument. Let c be as in (12), let $g \mapsto g'$ be the standard automorphism of G.

It follows from (12) that

$$(dF)_{gx}(g\,y) = c(g)(dF)_x(y),$$

whence, using (14),

$$B(g\,x, g'\,y) = F(g'(j\,x))^{-1}(dF)_{g'(jx)}(g'\,y) = B(x, y),$$

for all $g \in G$. Moreover it follows from 1.14 that $B \neq 0$.

Replacing y by $t\,y$ in (13) ($t \in K^*$) we obtain

$$F(x)F(t\,j\,x + j\,y)F(y) = F(x + t\,y),$$

if x, y and $j\,x + j\,y$ are invertible. Differentiation shows that

$$F(x)(dF)_{jy}(j\,x)F(y) = (dF)_x(y),$$

whence, replacing x by $j\,x$ and using (14),

$$B(x, y) = B(y, x),$$

for (x, y) in a suitable nonempty open subset of $V \times V$. This establishes the symmetry of B. Since B is linear in its second argument it follows that B is bilinear. The invariance has already been proved. This establishes 1.24.

1.25 Direct sum of J-structures. Let $\mathscr{S} = (V, j, e)$ and $\mathscr{S}' = (V', j', e')$ be two J-structures. We define their direct sum $\mathscr{S}'' = (V'', j'', e'')$ as follows. Take $V'' = V \times V'$, define

$$j''(v, v') = (j\,v, j'\,v'),$$

$$e = (e, e').$$

It is easily seen that $\mathscr{S}''$ satisfies the axioms of a J-structure. We write $\mathscr{S}'' = \mathscr{S} \oplus \mathscr{S}'$. We have (in an obvious sense) properties like $P'_{\mathscr{S}''} = P_{\mathscr{S}} \oplus P_{\mathscr{S}'}$, $\sigma_{\mathscr{S}''} = \sigma_{\mathscr{S}} \oplus \sigma_{\mathscr{S}'}$.

1.26 Fields of definition. Assume that $\mathscr{S} = (V, j, e)$ is a J-structure over the subfield k of K (in the sense of 1.3). Then the norm N is defined over k. It follows from 1.18 that the standard symmetric bilinear form σ is defined over k. Moreover it is clear from (16) that P is de defined over k. We cannot assert that the structure group G is defined over k in the sense of [1, p. 46] (a criterion for this to be the case will be given in 4.10).

§1. J-structures

But the inner structure group is defined over k: this follows from the fact that P is so, using [3, 2.2, p. 106].

1.27 J-structures without identity element. We terminate this section by discussing a situation which is somewhat more general than that of 1.3, in that we dispense with the "identity element" e. We call *J-structure without identity element* a pair $\mathscr{S} = (V, j)$ of a finite dimensional vector space V and a birational map j of V such that the following holds:

(J1)' *j is homogeneous of degree -1.*

If j is regular in x define $P(x)$ by (16), viz.

$$P(x) = -(dj)_x^{-1}.$$

We then further require the following identity of rational maps:

(J2)' $$jx = j(x+y) + j(x + P(x)jy),$$

and an analogue of (J3):

(J3)' *The structure group G of j has a dense orbit in V.*

Replace y by ty in (J2)' ($t \in K^*$). Comparing coefficients in a formal power series development, as in the proof of 1.16, one sees that

$$(dj)_x(y) + j(P(x)jy) = 0,$$

whence

$$j(P(x)jy) = P(x)^{-1}y,$$

if j is regular in x and y. It follows that $P(x) \in G$ if j is regular in x.

Now take $e \in V$ such that e is in the dense orbit of G in V and that j is regular in e, and put

$$j'x = P(e)jx.$$

Since 1.16(iii) holds for j and P (this is solely a consequence of the homogeneity of j) we have $j'e = e$. Let $\mathscr{S}' = (V, j', e)$.
One then checks without trouble that $\mathscr{S}'$ is a J-structure in the sense of 1.3. Thus we have reduced J-structures without identity to J-structures. It is easily seen that $\mathscr{S}'$ is uniquely determined by $\mathscr{S}$, up to isomorphism. We finally remark that (J2)' holds in any J-structure. This follows from (J2), using (J3) and 1.16(i), (iv).

Notes

Most of the material of §1 was suggested by results in the theory of Jordan algebras. The structure group of a Jordan algebra was first introduced by Koecher in [16, p. 70], in a somewhat different manner. See also [8, p. 79] and [17].
The axioms of a J-structure formalize the notion of inverse. (J1) and (J2) are then obvious requirements. The importance of (J3) was first realized by Braun and Koecher, see [8,

p. 152], where it is shown that properties of this nature can be used to characterize Jordan algebras, if the characteristic is not 2. We shall discuss this in §6.

Our definition of the norm in 1.4 does not reflect the current definition of norms in Jordan algebras, due to Jacobson, see [14, p. 223]. The procedure followed here is quite natural in the context of J-structures. It leads to simple proofs of results like 1.10 and 1.12, which are analogues of results contained in [8, III, §3]. 1.14 was also suggested by material from that book [loc. cit., III, §4].

The introduction of P by (16), in the case of Jordan algebras, is due to Koecher [16]. 1.16 reflects well-known properties in Jordan algebras. See also §3.

The inner structure group was introduced in [8, p. 92]. The name is due to Jacobson [14, p. 59]. The discussion of invariant bilinear forms was inspired by [8, III, §4].

The definition of a J-structure without identity in 1.27 was suggested by the notion of an isotope in a quadratic Jordan algebra, see [15, 1.63]. We shall not use this notion for J-structures.

§ 2. Examples

In this section we discuss some examples of J-structures. Almost all of them are related to associative algebras and quadratic forms.

2.1 Theorem. *Let A be a finite dimensional associative algebra with identity element e.*
(i) *There exists a unique birational map $i: A \to A$ with the following property: i is regular in x if and only if x has an inverse x^{-1}, and in that case we have $ix = x^{-1}$;*
(ii) $\mathscr{J}(A) = (A, i, e)$ *is a J-structure. If A is defined over k, then $\mathscr{J}(A)$ is defined over k.*

If $x \in A$ denote by $L(x)$ the linear transformation $y \mapsto x \cdot y$ of A. The associativity of A implies that $L(x) L(y) = L(xy)$; clearly $L(e) = \mathrm{id}$. If x has an inverse, then $L(x) L(x^{-1}) = \mathrm{id}$, so that $L(x)$ is a nonsingular linear transformation. Conversely if $L(x)$ is nonsingular, then x has an inverse, namely $L(x)^{-1} e$. Since $L(x)$ depends linearly on x and since the inverse of a nonsingular linear transformation of A is a rational function of that linear transformation, we conclude that

$$ix = L(x)^{-1} e$$

defines a rational map $A \to A$, which has the property of (i). The uniqueness of i follows from the uniqueness of the inverse of an element of A. Let A^* be the set of invertible elements of the associative algebra A. This is an open subset of A.

To prove (ii) we have to check the axioms (J 1), (J 2), (J 3) of 1.3 for (A, i, e). For (J 1) and (J 2) this is easy, and the details are omitted (for (J 2) use that $(e + x^{-1})^{-1} = x(e+x)^{-1}$ if x and $e+x$ are invertible).

To establish (J 3) we observe that for $x, a, b \in A^*$ we have $(axb)^{-1} = b^{-1} x^{-1} a^{-1}$, or $i(abx) = b^{-1}(ix) a^{-1}$. This shows that all linear transformations

$$x \mapsto a x b \, (a, b \in A^*)$$

are in the structure group G of i. It follows that Ge contains the nonempty open subset A^* of A, which proves (J 3). The last point of (ii) is clear, in view of the definitions.

We call $\mathscr{J}(A)$ *the J-structure defined by the associative algebra A*. Clearly $x \in A$ is invertible in $\mathscr{J}(A)$ if and only if x is invertible in the associative algebra A.

2.2. We next determine the rational map $P: A^* \to GL(A)$ of the J-structure $\mathscr{J}(A)$ (see 1.15). For $x, y \in A^*$ such that $x + y \in A^*$ we have
$$(x+y)^{-1} = x^{-1}(e + y x^{-1})^{-1},$$
or
$$i(x+y) = L(x^{-1}) \cdot i(e + y x^{-1}).$$
This implies that
$$(di)_x(y) = L(x^{-1})(di)_e(y x^{-1}).$$
By 2.1 and 1.16(iv) we have $(di)_e(z) = -z$, whence
$$(di)_x(y) = -x^{-1} y x^{-1},$$
$$P(x) y = x y x.$$
Clearly P can be extended to a quadratic map $A \to \operatorname{End}(A)$.

2.3. We call the norm N (the trace τ) of $\mathscr{J}(A)$ the *reduced norm* (respectively, the *reduced trace*) of A. We shall show presently that these notions are the usual ones. Let σ be the standard symmetric bilinear form on $A \times A$ (see 1.4 and 1.11 for N, σ and τ).

2.4 Lemma. *If $x, y \in A$ then $N(xy) = N(x) N(y)$, $\sigma(x,y) = \tau(xy)$.*

From the proof of axiom (J 3) in 2.1 we conclude that if $x \in A^*$ the linear transformation $L(x)$ lies in the structure group G of $\mathscr{J}(A)$. It then follows from 2.1 and 1.5 that there exists $c \in K^*$ such that for fixed $x \in A$ we have
$$N(xy) = c N(y) \quad (y \in A).$$
Taking $y = e$ we obtain the first formula for $x \in A^*$, $y \in A$, hence for all x, y.
We have
$$N(x) N(ix + y) = N(x) N(x^{-1} + y) = N(e + xy)$$
$$= 1 + (dN)_e(xy) + \text{terms of degree} \geq 2.$$
The second formula of 2.4 now follows from the definitions of σ and τ (given in 1.11), observing that $\tau(x) = (dN)_e(x)$.

2.5. Now let $A = \mathbb{M}_r$, the algebra of $r \times r$ matrices. For $X = (x_{ij}) \in \mathbb{M}_r$, let X_{ij} be the determinant of the $(r-1) \times (r-1)$ matrix, obtained by deleting the i-th row and the j-th column of X, multiplied by $(-1)^{i+j}$. Put $\operatorname{adj}(X) = (X_{ji})$ ($1 \leq i, j \leq r$). Then if X is an invertible matrix we have, as

§2. Examples

is well-known

(1) $$X^{-1} = (\det X)^{-1} \cdot \operatorname{adj}(X).$$

From the formula

(2) $$\det X = \sum_{j=1}^{r} x_{ij} X_{ji}$$

one easily deduces the well-known fact that the polynomial map $\det \colon \mathbb{M}_r \to \mathbb{M}_r$ is irreducible (use induction on r).

It follows that adj and det are a numerator and a denominator of i, respectively. Since $\det e = 1$, we conclude that det is the norm of the J-structure $\mathscr{J}(\mathbb{M}_r)$. Hence $\mathscr{J}(\mathbb{M}_r)$ has degree r.

It also follows that in $\mathbb{M}_r$ the norm coincides with the reduced norm, which justifies our definition of reduced norm in an associative algebra (for the usual definition of reduced norm in a simple algebra see [6, p. 173]).

Also (tr denoting the trace function on $\mathbb{M}_r$)

$$\det(e+X) = 1 + \operatorname{tr}(X) + \text{terms of degree} \geq 2,$$

which shows that
$$\tau(X) = \operatorname{tr}(X).$$

If A is a central simple algebra over a field $k \subset K$, the norm N and the trace τ are defined over k. By loc. cit. $N(x)(\tau(x))$ is the reduced norm of $x \in A(k)$ (respectively, the reduced trace of x). We denote by $\mathscr{M}_r$ the J-structure $\mathscr{J}(\mathbb{M}_r)$ of degree r, discussed above. From what was said above it follows that the norm of $\mathscr{M}_r$ is an irreducible polynomial function. Also, the standard symmetric bilinear function is nondegenerate in this case. Hence N is a nondegenerate polynomial function.

The next result will only be needed in Section 15, in the proof of 15.8. If A is an associative algebra we denote by A° the opposite algebra, whose underlying vectorspace is that of A, with the product in reversed order. Let d be the diagonal imbedding $a \mapsto (a, a)$ of A into $A \oplus A^\circ$.

2.6 Theorem. *Let $A = \mathbb{M}_r$. Let ϕ be a linear map of A into an associative algebra B with $\phi(e) = e$. Assume that $\phi(x)$ is invertible in B if x is so in A, with $\phi(x)^{-1} = \phi(x^{-1})$. Then there exists a unique algebra homomorphism ψ of $A \oplus A^\circ$ into B such that, d being as above, $\phi = \psi \circ d$. If $k \subset K$ is a subfield such that B and ϕ are defined over k, then ψ is defined over k.*

Let i_A, i_B be as in 2.1, for A and B. Our assumption about ϕ means that $i_B \circ \phi = \phi \circ i_A$. The chain rule for differentials 0.11 then shows that we have, if $x \in A^*$

$$(d i_B)_{\phi(x)} \circ \phi = \phi \circ (d i_A)_x.$$

Using 2.2 one concludes that

(3) $$\phi(x y x) = \phi(x) \phi(y) \phi(x) \quad (x, y \in A)$$

whence

(4) $$\phi(xy+yx)=\phi(x)\phi(y)+\phi(y)\phi(x)$$

(replace in (3) y by e and x by $x+y$).

Let (E_{ij}) be the canonical basis of $\mathbb{M}_r$. Put $\phi E_{ij}=b_{ij}$. If $x_i\in K^*$, $x\in K$, $i\ne j$, we have

$$(x_1 E_{11}+\cdots+x_r E_{rr}+x E_{ij})^{-1}=x_1^{-1} E_{11}+\cdots+x_r^{-1} E_{rr}-x x_i^{-1} x_j^{-1} E_{ij},$$

which implies a similar formula, with E_{rs} replaced by b_{rs}. From this one concludes that the b_{ii} are mutually orthogonal idempotents in B, with sum e and that

(5) $$b_{ii}b_{ij}=b_{ij}b_{jj}, \quad b_{jj}b_{ij}=b_{ij}b_{ii}, \quad b_{ij}b_{kk}=b_{kk}b_{ij}=0,$$

if i,j,k are distinct. If $i\ne j$ put

$$e_{ij}=b_{ii}b_{ij}, \quad f_{ij}=b_{jj}b_{ij}.$$

It follows that
$$b_{ij}=e_{ij}+f_{ij}$$
and that

(6) $$\begin{cases} e_{ij}=b_{ii}b_{ij}=b_{ij}b_{jj}=(e-b_{jj})b_{ij}=b_{ij}(e-b_{ii}) \\ f_{ij}=b_{jj}b_{ij}=b_{ij}b_{ii}=(e-b_{ii})b_{ij}=b_{ij}(e-b_{jj}). \end{cases}$$

We claim that $e_{ij}f_{kl}=f_{kl}e_{ij}=0$. For by (5) and (6)

$$e_{ij}f_{kl}=b_{ij}b_{jj}b_{kl}b_{kk}=0 \qquad \text{if } j\ne k,\ j\ne l,$$
$$=b_{ii}b_{ij}b_{ll}b_{kl}=0 \qquad \text{if } l\ne i,\ l\ne j,$$
$$=b_{ij}b_{jj}(e-b_{jj})b_{ji}=0 \qquad \text{if } j=k,\ l=i.$$

In the remaining case $l=j$ we have, using (4),

$$e_{ij}f_{kj}=b_{ii}b_{ij}b_{kj}b_{kk}=-b_{ii}b_{kj}b_{ij}b_{kk}=0 \qquad \text{if } i\ne k,$$

and
$$e_{ij}^2=b_{ii}b_{ij}^2 b_{ii}=0,$$

by (3). That $f_{kl}e_{ij}=0$ follows similarly.

One then proves, by an argument of the same kind that

(7) $$\begin{cases} e_{ij}e_{kl}=\delta_{jk}e_{il} & \text{if } i\ne l \\ f_{ij}f_{kl}=\delta_{il}f_{kj} & \text{if } j\ne k. \end{cases}$$

We next claim that $e_{ij}e_{ji}$ is independent of $j\ne i$. Let i,j,k be distinct. By (7) we have $e_{ij}e_{jk}=e_{ik}$, $e_{ki}=e_{kj}e_{ji}$, whence $e_{ik}e_{ki}=e_{ij}e_{jk}e_{kj}e_{ji}$. Now (4) implies that

$$b_{jk}b_{kj}+b_{kj}b_{jk}=b_{jj}+b_{kk},$$

whence, multiplying by b_{jj} on both sides,

$$e_{jk}e_{kj}+f_{kj}f_{jk}=b_{jj}.$$

§2. Examples

It then follows that

$$e_{ik}\, e_{ki} = e_{ij}(b_{jj} - f_{kj}\, f_{ik})\, e_{ji} = e_{ij}\, b_{jj}\, e_{ji} = e_{ij}\, e_{ji},$$

which establishes our claim. Similarly, $f_{ji}\, f_{ij}$ is independent of $j \neq i$. Put $e_{ii} = e_{ij}\, e_{ji}$, $f_{ii} = f_{ji}\, f_{ij}$. Then $e_{ii} + f_{ii} = b_{ii}$.

One now shows that (7) holds without restriction. Defining

$$\psi(E_{ij}, E_{kl}) = e_{ij} + f_{kl}$$

it follows that ψ has the required properties.

To prove the uniqueness of ψ remark that

$$d(E_{ij})\, d(E_{kl}) = (\delta_{jk}\, E_{il},\, \delta_{il}\, E_{kj}),$$

from which one concludes that ψ must satisfy

$$\delta_{jk}\, \psi(E_{il}, 0) + \delta_{il}\, \psi(0, E_{kj}) = b_{ij}\, b_{kl}.$$

It follows that

$$\psi(E_{ij}, 0) = b_{ij}\, b_{jj} = e_{ij},$$
$$\psi(0, E_{ij}) = b_{ij}\, b_{ii} = f_{ij},$$

establishing uniqueness. The last point of the assertion is clear.

2.7 Algebras with involution. Now let A be a finite dimensional associative algebra with identity e, let ρ be an involution of A. This means that ρ is a linear bijection of A onto A such that $\rho^2 = \text{id}$ and that $\rho(x\, y) = \rho(y)\, \rho(x)$ ($x, y \in A$).

Let

$$B = \{x \in A \mid \rho x = x\}$$

be the space of ρ-invariant elements of A. Put $B^* = B \cap A^*$, the set of invertible elements of B. Since $\rho(x)^{-1} = \rho(x^{-1})$, i induces a rational map $B \to B$, which we also denote by i.

2.8 Proposition. *(B, i, e) is a J-structure if the following condition is satisfied: the subset $\{x \cdot \rho\, x \mid x \in A\}$ of B contains a nonempty Zariski-open subset of B. This condition holds if $\text{char}(K) \neq 2$.*

It follows from 2.1 that the axioms (J1) and (J2) of 1.3 hold in (B, i, e). It remains to prove (J3). Now if $a \in A^*$, $x \in B^*$, we have $a\, x \cdot \rho\, a \in B^*$, and

$$i(a\, x \cdot \rho\, a) = (\rho\, a)^{-1}(i\, x)\, a^{-1}.$$

It follows that for $x \in A^*$ the linear transformation

$$y \mapsto x\, y \cdot \rho\, x \qquad (y \in B)$$

of B lies in the structure group G of the restriction of i to B.

Also, taking $x \in B^*$, we conclude that $x^2 = x e \cdot (\rho x)$ lies in Ge. That (J 3) holds if the condition of 2.8 holds is now clear. The last assertion follows from the following lemma.

2.9 Lemma. *Let* $\mathrm{char}(K) \neq 2$. *Let A be a not necessarily associative algebra with identity e, let B be a subspace of A, containing e and such that the square of an element of B lies in B. Then the set of squares of elements of B is dense in B.*

Proof of the lemma. Let ϕ be the morphism $x \mapsto x^2$ of B into B, then $\phi(e) = e$. The differential $(d\phi)_e$, which is a linear transformation of B, is scalar multiplication by 2. Since $\mathrm{char}(K) \neq 2$, it then follows that $\phi(B)$ is dense in B (see [1, 17.3, p. 75]). This concludes the proof of 2.7.

We write $(B, i, e) = \mathscr{J}(A, \rho)$ and we call it *the J-structure defined by the associative algebra with involution* (A, ρ), if the condition of 2.8 holds.

We next discuss some particular cases of 2.8.

2.10 Symmetric matrices. Take in 2.7 for A the algebra $\mathbb{M}_r$ of $r \times r$ matrices and let $\rho X = {}^t X$, the transposed of X. Then B is the set $\mathbb{S}_r$ of symmetric $r \times r$ matrices. By 2.8 we obtain a J-structure $(\mathbb{M}_r, \rho)$, if $\mathrm{char}(K) \neq 2$. As a matter of fact this restriction on $\mathrm{char}(K)$ is superfluous, so that $\mathscr{S}_r = (\mathbb{S}_r, i, e)$ is always a J-structure. In order to establish this it suffices, by 2.8, to show that the set of symmetric matrices of the form $X \cdot {}^t X$ ($x \in \mathbb{M}_r^*$) is dense in $\mathbb{S}_r$. But this is a consequence of the fact that in any characteristic two nondegenerate symmetric bilinear forms on $K^r \times K^r$ are equivalent, which follows from the results of [12, p. 15] and [12, p. 20].

Now the restriction of the polynomial function $\det$ of $\mathbb{M}_r$ to $\mathbb{S}_r$ is irreducible (this can be proved by using (2) in the same way as the irreducibility of the form $\det$ on $\mathbb{M}_r$, see 2.5). It follows as in 2.5 that $\det$ is the norm of the J-structure $\mathscr{S}_r$, hence $\mathscr{S}_r$ is a J-structure of degree r.

From 2.4 we conclude that the standard bilinear symmetric form σ on $\mathscr{S}_r$ is again given by
$$\sigma(X, Y) = \mathrm{tr}(XY).$$
Let $X = (x_{ij})$, $Y = (y_{ij})$ ($x_{ij} = x_{ij}$, $y_{ij} = y_{ji}$). Then
$$\sigma(X, Y) = \sum_{i=1}^{r} x_{ii} y_{ii} + 2 \sum_{1 \leq i < j \leq r} x_{ij} y_{ij}.$$
It follows that for $r > 1$, σ is nondegenerate if and only if $\mathrm{char}(K) \neq 2$ (if $r = 1$, it is clear that $\mathscr{S}_r = \mathscr{M}_r$). Hence the norm of $\mathscr{S}_r$ is a nondegenerate polynomial function if $\mathrm{char}(K) \neq 2$.

2.11 Alternating matrices. Let $\mathbb{A}_r$ be the set of alternating $r \times r$ matrices, i.e. $\mathbb{A}_r$ consists of the $X = (x_{ij})$ in $\mathbb{M}_r$, such that $x_{ii} = 0$, $x_{ij} = -x_{ji}$ ($1 \leq i, j \leq r$).

§2. Examples

2.12 Lemma. (i) *There exists a homogeneous polynomial function* Pf *on* $\mathbb{A}_{2r}$ *of degree r, such that for* $X \in \mathbb{A}_{2r}$ *we have that* $\det X = (\operatorname{Pf} X)^2$;
(ii) *There exists a polynomial map* $n: \mathbb{A}_{2r} \to \mathbb{A}_{2r}$ *of degree* $r-1$, *such that for nonsingular* $X \in \mathbb{A}_{2r}$ *we have*

$$X^{-1} = (\operatorname{Pf} X)^{-1} n(X);$$

(iii) *The polynomial function* Pf *is irreducible.*

(Pf(X) is called the Pfaffian of X.) (i) and (ii) are well-known results, a proof can be found in [14, Lemma 2, p. 230]. The proof of (iii) uses the analogue of (2) for Pf, and is similar to the irreducibility proof for det.

2.13. Let $S = (s_{ij}) \in \mathbb{A}_{2r}$ be given by $s_{ij} = 0$ if $|i-j| \neq 1$, $s_{i,i+1} = 1$, $s_{i+1,i} = -1$. S is nonsingular and $S^{-1} = S$. Let now ρ denote the involution

$$X \mapsto S \cdot {}^t X \cdot S^{-1}$$

of $\mathbb{M}_{2r}$. As in 2.7, let $B = \{x \in \mathbb{M}_{2r} | \rho X = X\}$.

By 2.8, $(B, i, e) = \mathscr{J}(\mathbb{M}_{2r}, \rho)$ is a J-structure if $\operatorname{char}(K) \neq 2$. Then $X \mapsto XS$ defines a linear isomorphism of B onto $\mathbb{A}_{2r}$, by means of which one can transport the J-structure $\mathscr{J}(\mathbb{M}_{2r}, \rho)$ to a J-structure $\mathscr{A}_r$ with underlying vector space $\mathbb{A}_{2r}$. One can define $\mathscr{A}_r$ directly as follows: $\mathscr{A}_r = (\mathbb{A}_{2r}, j, S)$, where j is given by

$$jX = SX^{-1} \cdot S^{-1},$$

if $X \in \mathbb{A}_{2r}$ is nonsingular.

We started with the restriction $\operatorname{char}(K) \neq 2$. However $\mathscr{A}_r$ is a J-structure in all characteristics: the crucial axiom is (J 3), which holds in $\mathscr{A}_r$ because any two nondegenerate alternating bilinear forms on $K^{2r} \times K^{2r}$ are equivalent (see [12, p. 15]).

From 2.12 we conclude that $\mathscr{A}_r$ is a J-structure of degree r and that Pf is the norm of $\mathscr{A}_r$ (if Pf is normalized such that Pf$(S) = 1$). By 2.12(iii) the norm is irreducible. The standard symmetric bilinear form σ is given by the following result.

2.14 Lemma. *Let* $X = (x_{ij})$, $Y \in (y_{ij})$ *be matrices in* $\mathbb{A}_{2r}$. *Then*

$$\sigma(X, Y) = - \sum_{1 \leq i < j \leq 2r} x_{ij} y_{ij}.$$

If X is nonsingular we have (see 1.11)

$$\Phi(X, Y) = \operatorname{Pf}(X) \operatorname{Pf}(X^{-1} + Y) = 1 + \sigma(X, Y) + \text{terms of degree} \geq 2.$$

By 2.12(i) we have

(8)
$$\Phi(X, Y)^2 = \det X \cdot \det(X^{-1} + Y) = \det(e + XY)$$
$$= 1 + \operatorname{tr}(XY) + \text{terms of degree} \geq 2.$$

First let char$(K) = 0$. From the last two formulas it then follows that
$$\sigma(X, Y) = \tfrac{1}{2}\operatorname{tr}(XY) = -\sum_{1 \leq i < j \leq 2r} x_{ij}\, y_{ij}.$$
Now from (8) it follows that $\Phi(X, Y)$ is a polynomial function with *integral* coefficients. But then the validity of 2.14 in characteristic 0 implies its truth in any characteristic. Observe that σ is nondegenerate in all characteristics.

2.15 Quadratic forms. Let V be a finite dimensional vector space, let Q be a quadratic form on V (see 0.16). We put
$$Q(x, y) = Q(x+y) - Q(x) - Q(y),$$
so $(x, y) \mapsto Q(x, y)$ is a symmetric bilinear form on $V \times V$. We put
$$Q_x(y) = Q(x, y),$$
this is a linear function on V.

Let $e \in V$ be such that $Q(e) = 1$. We put
$$\bar{x} = -x + Q(x, e)\, e.$$
Since $Q(e, e) = 2Q(e) = 2$, we have $\bar{e} = e$. It is easily seen that
$$(9) \qquad\qquad Q(\bar{x}) = Q(x),$$
whence
$$(10) \qquad\qquad \bar{\bar{x}} = x.$$

Let G be the group of all linear transformations $g \in GL(V)$ such that there exists $a \in K^*$ with
$$Q(g\,x) = a\,Q(x) \qquad (x \in V).$$
If $g \in G$, we put
$$(11) \qquad\qquad g' \cdot x = a^{-1}\, \overline{g(\bar{x})}.$$
Then $g' \in G$ and $g \mapsto g'$ is an automorphism of G of order 2.

We need the following lemma, which can be viewed as a version of Witt's theorem (see [12, p. 21]).

2.16 Lemma. *Let $x, y \in V$, $Q(x)\,Q(y) \neq 0$. There exists $g \in G$ such that $gx = y$ if one of the following conditions holds:*
(a) char$(K) \neq 2$,
(b) $Q_x \neq 0$, $Q_y \neq 0$.

Since scalar multiplications $x \mapsto t\,x$ $(t \in K^*)$ are in G, we may assume that $Q(x) = Q(y)$. Now
$$Q(x+y) + Q(x-y) = 2Q(x) + 2Q(y) = 4Q(x).$$

§2. Examples

If $\mathrm{char}(K) \neq 2$ it follows that at least one of the two elements $Q(x+y)$, $Q(x-y)$ (say the first one) is nonzero. Let r_{x+y} be the reflection defined by $x+y$, i.e.
$$r_{x+y}(z) = z - Q(x+y)^{-1} Q(x+y, z)(x+y).$$
Since
$$Q(x+y, x) = 2Q(x) + Q(x, y) = Q(x+y),$$
we have
$$r_{x+y}(x) = -y,$$
so $g = -r_{x+y}$ (which is in G) is as required.

If $\mathrm{char}(K) = 2$ and if (b) holds there exists $a \in V$ such that $Q(a) \neq 0$, $Q(x,a) \neq 0$, $Q(y,a) \neq 0$. We may then assume that $Q(a) = Q(x) = Q(y)$. But then
$$Q(a+x) = Q(a) + Q(x) + Q(a, x) = Q(a, x) \neq 0.$$
As before we see that $r_{a+x}(x) = a$, similarly $r_{a+y}(y) = a$. Hence
$$r_{a+y}^{-1} r_{a+x}(x) = y,$$
which implies the assertion.

With the above notations define for $x \in V$, $Q(x) \neq 0$,
$$jx = Q(x)^{-1} \cdot \bar{x}.$$
Clearly j is a rational map $V \to V$.

2.17 Theorem. *Assume that $Q_e \neq 0$. Then $\mathscr{J}(V, Q, e) = (V, j, e)$ is a J-structure.*

From (9) and (10) we obtain
$$Q(jx) = Q(x)^{-2} Q(\bar{x}) = Q(x)^{-1},$$
and
$$j(jx) = Q(jx)^{-1} j\bar{x} = \bar{\bar{x}} = x.$$
Also $je = Q(e)^{-1} \bar{e} = e$. Hence (J1) holds.

To prove (J2), observe first that (if $Q(x) \neq 0$)
$$Q(e + jx) = Q(e + Q(x)^{-1}\bar{x}) = Q(x)^{-2} Q(Q(x)e + \bar{x})$$
$$= Q(x)^{-2} \{Q(x)^2 + Q(x)Q(x,e) + Q(x)\} = Q(x)^{-1} Q(e+x).$$
Hence (if $Q(x) \neq 0$, $Q(e+x) \neq 0$)
$$j(e+x) + j(e+jx) = Q(e+x)^{-1} \overline{e+x} + Q(x) Q(e+x)^{-1} \overline{e+jx}$$
$$= Q(e+x)^{-1} \overline{e+x+Q(x)e+\bar{x}} = e,$$
which proves (J2).

To prove (J3), observe that for $g \in G$, and g' being given by (11), we have
$$g(jx) = j(g'x).$$

Hence G is contained in the structure group of j. By 2.16 we have that the orbit Ge contains the nonempty open set of all $x \in V$ such that $Q(x) \neq 0$, $Q_x \neq 0$. Hence (J 3) holds.

2.18 Corollary. *G is the structure group of $\mathscr{J}(V, Q, e)$.*

This follows by using 1.5.
In the situation of 2.17 we call $\mathscr{J}(V, Q, e)$ a *J-structure defined by the quadratic form* Q.

Remark. The hypothesis $Q_e \neq 0$ of 2.17 is automatically fulfilled if $\operatorname{char}(K) \neq 2$, since we have $Q_e(e) = Q(e, e) = 2$.
We next shall determine norm and standard symmetric bilinear form. First we establish a simple lemma.

2.19 Lemma. *Let $\mathscr{S} = (V, j, e)$ be a J-structure of degree 1. Then $\dim V = 1$ and $\mathscr{S}$ is isomorphic to $\mathscr{J}(K)$, the J-structure defined by the K-algebra K.*

If S has degree 1, the numerator of n of j must be a constant polynomial map $V \to V$. It follows that $j(V)$ has dimension 1. Since j is birational this implies that $\dim V = 1$. The assertion then follows.

Returning to $\mathscr{J}(V, Q, e)$ assume from now on that $\dim V > 1$. It then follows from 2.19 that the polynomial function Q is the norm of $\mathscr{J}(V, Q, e)$.
Since

$$Q(x) Q(jx + y) = Q(x) Q(Q(x)^{-1} \bar{x} + y)$$
$$= Q(x) \{Q(x)^{-2} Q(\bar{x}) + Q(x)^{-1} Q(\bar{x}, y) + Q(y)\}$$
$$= 1 + Q(\bar{x}, y) + Q(x) Q(y),$$

we conclude from 1.11 that

(12) $$\sigma(x, y) = Q(\bar{x}, y).$$

Hence σ is nondegenerate if and only if the quadratic form Q is nondegenerate if $\operatorname{char}(K) \neq 2$ and is nondefective if $\operatorname{char}(K) = 2$ (for these notions see 0.16).
From $Q(x) jx = \bar{x}$ one obtains that

$$Q(x)(dj)_x(y) + (dQ)_x(y) jx = \bar{y}.$$

Since
$$(dQ)_x(y) = Q(x, y),$$
this gives
$$(dj)_x(y) = Q(x)^{-1} \bar{y} - Q(x)^{-1} Q(x, y) jx$$
$$= Q(jx) \bar{y} - Q(jx, \bar{y}) jx.$$

§2. Examples

Hence

(13) $$P(x)\,y = Q(x,\bar{y})\,x - Q(x)\,\bar{y}.$$

Finally, suppose that V has a k-structure, that Q is defined over k and that $e \in V(k)$. Then it is clear that $\mathscr{J}(V, Q, e)$ is defined over k.

2.20. Let $\operatorname{char}(K) \neq 2$. Let $(e_i)_{1 \leq i \leq r}$ be the canonical basis of K^r, let Q_r be the quadratic form on K^r defined by

$$Q_r\left(\sum_{i=1}^{r} x_i\, e_i\right) = \sum_{i=1}^{r} x_i^2.$$

If $r > 1$ then $\mathscr{J}(K^r, Q_r, e_r)$ is a J-structure of degree 2, which we denote by $\mathcal{O}_{2,r}$.

Now assume $\operatorname{char}(K) = 2$. Let Q'_r be the quadratic form on K^{2r} defined by

$$Q'_r\left(\sum_{i=1}^{2r} x_i\, e_i\right) = \sum_{i=1}^{r} x_i\, x_{i+r}.$$

We denote the J-structure $\mathscr{J}(K^{2r}, Q'_r, e_r + e_{2r})$ by $\mathcal{O}'_{2,r}$ ($r \geq 1$). Next let $(e_i)_{0 \leq i \leq 2r}$ be the canonical basis of K^{2r+1} ($r \geq 1$). Let Q''_r be the quadratic form on K^{2r+1} defined by

$$Q''_r\left(\sum_{i=0}^{2r} x_i\, e_i\right) = x_0^2 + \sum_{i=1}^{r} x_i\, x_{i+r}.$$

We denote the J-structure $\mathscr{J}(V, Q''_r, e_r + e_{2r})$ by $\mathcal{O}''_{2,r}$. The J-structures $\mathcal{O}'_{2,r}$ and $\mathcal{O}''_{2,r}$ are only defined if $\operatorname{char}(K) = 2$.

Using (12) one easily checks that $\mathcal{O}_{2,r}$ and $\mathcal{O}'_{2,r}$ have a nondegenerate standard bilinear form σ. This is not so for $\mathcal{O}''_{2,r}$. On sees that in that case $\sigma(x, V) = 0$ if and only if $x \in K\, e_0$.

2.21. We finally give a brief discussion of a more recondite example, of a quite different nature. This example will not be used in the sequel. We use freely results and notions from the theory of linear algebraic groups, for which we refer to [1].

Let G be a linear algebraic group which is connected and reductive. Let $\mathfrak{g}$ be its Lie algebra. G and its subgroups act on $\mathfrak{g}$ via the adjoint representation Ad. Assume that S is a 1-dimensional subtorus of G with the following properties:

(A) S has only two nonzero weights a and $-a$ in $\mathfrak{g}$;
(B) there exists $n \in G$ such that $n s n^{-1} = s^{-1}$ ($s \in S$).

Denote by H the centralizer of S in G. Then H is a connected reductive subgroup of G. $\mathfrak{h}$ denoting its Lie algebra, we have a decomposition

$$\mathfrak{g} = \mathfrak{h} \oplus \mathfrak{g}_a \oplus \mathfrak{g}_{-a},$$

where
$$\mathfrak{g}_{\pm a} = \{X \in \mathfrak{g} \mid \mathrm{Ad}(s) X = s^{\pm a} X, \text{ for all } s \in S\}.$$

Then $\mathfrak{h} \oplus \mathfrak{g}_a$ and $\mathfrak{h} \oplus \mathfrak{g}_{-a}$ are the Lie algebras of parabolic subgroups Q, Q^-, respectively, see [4, 4.15, p. 90–91]. The unipotent radicals V and V^- of Q, Q^- have the Lie algebras $\mathfrak{g}_a, \mathfrak{g}_{-a}$, respectively. We have $S \subset H$ and these two groups act on V by conjugation. From [4, 3.17, p. 81] it follows that V is a vector group, and that the scalar multiplications of the vector group structure are given by conjugation with suitable elements of S.

Let n be as in (B) above. Then $Q^- = n Q n^{-1}$, $V^- = n V n^{-1}$, and n normalizes H. We write $h' = n h n^{-1}$ ($h \in H$). Put $n^2 = h_0$, then $h_0 \in H$.

Since Q and Q^- are opposite parabolic groups in the sense of [4, p. 88], we have that $V^- \cdot Q$ is an open subset of G [loc. cit., 4.10 d), p. 89]. Likewise, VnQ is an open subset of G. It follows that there exists a birational map of $V^- \cdot P$ onto VnQ. From this one concludes that there exists a unique birational map j of V to V such that

(14) $$j x \cdot n x n^{-1} \in n Q,$$

if j is regular in $x \in V$. This implies that

$$j(h x h^{-1}) = h'(j x)(h')^{-1}$$

in particular

$$j(s x s^{-1}) = s^{-1}(j x) s \quad (s \in S),$$

which means that j is homogeneous of degree -1 (for the vector space structure of V).

(14) can also be reformulated as follows: there exist unique rational maps $j, j_1: V \to V$ and $\lambda: V \to H$ such that

$$j x \cdot n x n^{-1} \cdot j_1 x = n \lambda(x).$$

j_1 is again homogeneous of degree -1. Replacing x by x^{-1} and taking inverses, one see that

$$j_1 x \cdot (n x n^{-1}) \in n Q,$$

whence $j_1 = j$. Hence

(15) $$j x \cdot n x n^{-1} \cdot j x = n \lambda(x).$$

We now also find that

(16) $$\lambda(h x h^{-1}) = h \lambda(x)(h')^{-1}.$$

It follows from (15) that

$$j(x y) j(x)^{-1} n \bigl(\lambda(x) j(x)^{-1} j(y)^{-1} \lambda(x)^{-1} \bigr) n^{-1} \in n Q,$$

§2. Examples

whence

(17) $$j(xy) \cdot j(\lambda(x)j(x)j(y)\lambda(x)^{-1}) = jx.$$

We now change the notation. Write the group operation in V additively, and put
$$P(x)y = \lambda(x)y\lambda(x)^{-1},$$
if λ is regular in x. (16) implies that $x \mapsto P(x)$ is a rational map of V into End(V) which is homogeneous of degree 2. (17) can then be rewritten as
$$j(x+y) + j(P(x)(jx+jy)) = jx.$$
Replacing x by tx ($t \in K^*$) and arguing as in the proof of 1.16 one sees that
$$P(x)jx = x, \quad P(x) = -(dj)_x^{-1}.$$
It follows that V and j satisfy axioms (J1)′ and (J2)′ of 1.27.

If one assumes moreover

(C) *H has an open orbit in V*,

then (V,j) is a J-structure without identity in the sense of 1.27, which can then be modified as in 1.27 to obtain a J-structure in V.

We next show how, given a simple J-structure, one can construct a reductive group G containing a 1-dimensional torus S such that (A), (B), (C) hold. In the construction we make use of Weil's theorem about enlarging an algebraic group germ to an algebraic group, which we first recall (see [32, p. 373], see also [11, exposé XVIII]).

2.22. Suppose that X is an algebraic variety, assume we are given a rational map $p: X \times X \to X$, denoted by $p(x,y) = xy$ and an involutorial birational map $i: X \to X$, denoted by $ix = x^{-1}$, such that the following relations hold (it being assumed that everything needed for these relations to make sense is defined)

(18) $$x(yz) = (xy)z,$$

(19) $$x^{-1}(xy) = (yx)x^{-1} = y.$$

We call a triple (X, p, i) satisfying these conditions an *algebraic group germ*. It is *defined over k* if X, p and i are so. Weil's theorem is now as follows.

2.23 Theorem. *Let (X, p, i) be an algebraic group germ. There exists an algebraic group G together with a morphism $f: X \to G$ such that*
(i) $f(xy) = f(x)f(y)$ *if xy is defined;*
(ii) f *is an isomorphism of X onto an open subset of G.*

If $k \in K$ is a field over which the group germ is defined then G and f can be taken to be defined over k. Moreover G is uniquely determined up to a k-isomorphism.

Now let $\mathscr{S} = (V, j, e)$ be a J-structure. Let H be the inner structure group of $\mathscr{S}$ (defined in 1.19). If $a \in V$ denote by t_a the translation $x \mapsto x + a$ of V. We consider the t_a and the elements of H as birational maps of V.

2.24 Lemma. *The map α of $H \times V \times V$ into the set of birational maps of V with $\alpha(h, a, b) = h \circ t_a \circ j \circ t_b$ is injective.*

It suffices to prove that $h \circ t_a \circ j \circ t_b = j$ implies $a = b = 0$, $h = \mathrm{id}$. Now this relation implies
$$h(x) + h(a) = j(jx - b),$$
for x in a suitable open subset of V. Replacing x by tx ($t \in K^*$) we see that
$$t h(x) + h(a) = t j(jx - tb),$$
which can only be if $a = 0$. Then
$$h(x) = j(jx - tb),$$
whence $h = \mathrm{id}$, $b = 0$.

2.25. Let $X = H \times V \times V$. If $h \mapsto h'$ is the standard automorphism of H, then
(20) $$j \circ h \circ j = h',$$
see 1.2. Moreover by (J2)' of 1.27 (which holds in $\mathscr{S}$, see the final remark of 1.27) we have
(21) $$j \circ t_a \circ j = t_{ja} \circ j \circ t_{-a} \circ (-P(a)).$$

(20) and (21) imply that if $x, y \in X$, the birational map $\alpha(x) \circ \alpha(y)$ (where α is as in 2.24) lies again in $\alpha(X)$, at least if (x, y) is in a suitable dense open subset U of $X \times X$. Hence it makes sense to define a map $p: U \to X$ by
$$p(x, y) = \alpha^{-1}(\alpha(x) \circ \alpha(y))$$
(20) and (21) imply that in fact p is a *rational* map $X \times X \to X$. Likewise one defines an involutorial birational map $i: X \to X$ by
$$i(x) = \alpha^{-1}(\alpha(x)^{-1}).$$

One checks by applying α that (18) and (19) hold. Hence (X, p, i) is an algebraic group germ. Let G and f be as in 2.23. Since, as one readily sees, p is regular in all points $((h, a, 0), (h', a', 0))$ it follows that $f(H \times V \times \{0\})$ is a closed subgroup of G, isomorphic to the semi-direct product of H and V. We identify H and V with the corresponding images in G. Let $n = f(-\mathrm{id}, 0, 0)$. Then G is generated by H, V and n. Let L be the maximal connected linear algebraic subgroup of G. By a theorem of Chevalley, L is a closed normal subgroup of G, and G/L is an abelian variety (for Chevalley's theorem see [25]). From what we just observed one infers that G/L is generated by the canonical image of n. This can only be if $G=L$, hence G is a connected *linear* algebraic group.

We have constructed G without any extra assumption on $\mathscr{S}$. Now assume moreover

(I) $\mathscr{S}$ is simple, i.e. *has no proper ideals* (see 1.3).
(II) H is reductive.
(We shall see in 14.27 that (II) is a consequence of (I).)

We show that (I) and (II) imply that G is reductive. Let $S \subset H$ be the 1-dimensional torus of scalar multiplications. S, considered as a subtorus of G, satisfies the conditions (A) and (B) of 2.21. Let R be the unipotent radical of G, put $\bar{G}=G/R$ and define $\bar{V}, \bar{H}, \bar{S}$ to be the canonical images of V, H, S in $\bar{G}$, respectively.

We first observe that $V \not\subset R$. For $V \subset R$ would imply, using (21), that we had $P(a)^{-1} P(b) \in R$ for all a, b in a suitable dense open subset of V, containing e. This implies $P(b) \in R$, hence $H \subset R$ and $G=R$, which is impossible.

n being as above, it follows from (21) that we have

(22) $$j a \cdot n x n^{-1} \in n H V.$$

The 1-dimensions subtorus $\bar{S}$ of the reductive group $\bar{G}$ has the properties (A) and (B) of 2.21. Proceeding as in 2.21, one defines a birational map $\bar{j}$ of $\bar{V}$ such that the analogue of (14) holds. Comparison with (22) then shows that we must have

$$j(x+r)-j(x) \in R \cap V,$$

for all $r \in R \cap V$. Hence $R \cap V$ is an ideal in $\mathscr{S}$ (see 1.3). Since $V \not\subset R$, it follows from assumption (I) that $R \cap V$ is reduced to the neutral element. The same is true of $R \cap n V n^{-1}$. Let $\mathfrak{g}$ be the Lie algebra of G, let $\mathfrak{g}_a, \mathfrak{g}_{-a}$ be as in 2.21. $\mathfrak{g}_a$ is the Lie algebra of V.

Let $\mathfrak{r}$ be the Lie algebra of R. Then $\mathfrak{r} \cap \mathfrak{g}_a = \{0\}$: if $\mathfrak{r} \cap \mathfrak{g}_a \neq \{0\}$ it would follow from [3, 9.8, p. 487] that $R \cap V$ contained a 1-dimensional subgroup. Likewise $\mathfrak{r} \cap \mathfrak{g}_{-a} = \{0\}$. But it then follows that S acts trivially on $\mathfrak{r}$, which implies that S centralizes R, hence $R \subset H$. Assumption (II) now shows that R is trivial, hence G is reductive.

In fact, G is even semisimple. For let T be a maximal torus of H, this is also one of G. The imbedding of H into $GL(V)$ defines an immersive representation $T \to GL(V)$. It follows that all nonidentity elements of T act nontrivially on V. The center of G is contained in T and acts trivially on V, hence is reduced to the neutral element. This establishes the semisimplicity of G. It is readily seen that G is adjoint.

It also follows that the procedure of 2.21, applied to the present G and S leads to the birational map j of the J-structure $\mathscr{S}$.

2.26 The Lie algebra of G. Let G be the group constructed in 2.25. Let $\mathfrak{h}$ be the Lie algebra of H. The Lie algebra $\mathfrak{g}$ of G can then be described as follows. We anticipate some results to be proved in the next sections. In 3.4 it is shown that P being defined by §1, (16), the map $x \mapsto P(x)$ is a quadratic polynomial map of V into $\text{End}(V)$, so that it makes sense to write $P(x, y) = P(x+y) - P(x) - P(y)$. In 4.3 we shall see that the linear transformation $L(a, b)$ with

$$L(a, b)\, x = P(x, a)\, b$$

lies in $\mathfrak{h}$. Let $X \mapsto X'$ be the automorphism of $\mathfrak{h}$ defined by the standard automorphism of H.

With these notations we can describe $\mathfrak{g}$: we have

$$\mathfrak{g} = \mathfrak{h} \oplus V \oplus V,$$

the Lie product being given by

$$[(X, a, b), (Y, c, d)] = ([X, Y] + L(a, d) - L(b, c), Xc - Ya, X'd - Y'b).$$

This description of $\mathfrak{g}$ follows from the construction of G given in 2.25. We omit the proof.

Notes

The examples of this section, except for the final one, discussed from 2.21 onwards, are of an elementary nature and need no further comment. 2.6 is a result about the representation theory of $\mathscr{M}_r$. We indicate the context in which it should then be viewed. Let $\mathscr{S} = (V, j, e)$ be a J-structure. Let T be the tensor algebra over V and denote by $C(\mathscr{S})$ the quotient of T by the twosided ideal generated by all elements $x \otimes jx - e$ (j regular in x). We call $C(\mathscr{S})$ the *Clifford algebra* of S (in view of the analogy of the definition with that of the Clifford algebra of a quadratic form). Then $C(\mathscr{S})$ is an associative algebra which is universal for homomorphisms of $\mathscr{S}$ into J-structures of the form $\mathscr{J}(A)$, (A associative) in an obvious sense. 2.6 implies that $C(\mathscr{M}_r) \simeq \mathbb{M}_r \oplus \mathbb{M}_r^\circ$. We shall not pursue further the representation theory of J-structures in this book. Such a theory ought to be modeled on the representation theory of Jordan algebras. The latter theory (characteristic not 2) has been extensively developed, mainly by N. Jacobson. For a comprehensive account see his book [14]. The analogue of the Clifford algebra $C(\mathscr{S})$ is called there "special universal envelope".

The example discussed in 2.21 was suggested by a remark of A. Borel. A situation like that of 2.21 is discussed from a different point of view in [4a, §5].

The construction of G in 2.25 is essentially due to Koecher [18], if the characteristic is not 2. The use of Weil's theorem 2.23 simplifies the construction. The construction of the Lie algebra of G, indicated in 2.26, was first given by Tits [31], and has been studied in detail by Koecher [19]. It follows from 2.25 that a simple J-structure leads to a reductive group G, containing a 1-dimensional torus S, which has property (A) of 2.21. This could possibly be used to study simple J-structures. We shall not do this, but follow instead in §12 and §13 another way to obtain the classification of simple J-structures. We shall exploit there a similar situation, where we have a reductive group G with a 1-dimensional subtorus S, such that S has but few weights in a given representation of G.

§3. The Quadratic Map of a J-structure

3.1. Let $\mathscr{S}=(V,j,e)$ be a J-structure of degree d. We denote the norm $\mathscr{S}$ by N, and we put for invertible $x \in V$

$$jx = N(x)^{-1} \cdot n(x),$$

where the numerator n of j is a homogeneous polynomial map $V \to V$ of degree $d-1$, and $n(e)=e$ (see 1.4). We denote by G the structure group of $\mathscr{S}$ and by $g \mapsto g'$ its standard automorphism. Let the character $a: G \to K^*$ be as in 1.5. By 1.5 and §1, (5) we then have

(1) $$n(g\,x) = a(g)\,g'(n\,x) \qquad (g \in G,\ x \in V).$$

In this section we shall study in more detail the rational map $P: V \to \operatorname{End}(V)$ introduced in 1.15. Recall that we have, if x is invertible

$$P(x) = -(dj)_x^{-1}.$$

We put for invertible $x \in V$ and $y \in V$

(2) $$\phi(x,y) = N(x)\,n(jx + y).$$

Since n is a polynomial map, ϕ is a rational map $V \times V \to V$, whose denominator depends only on the first variable. Hence, for fixed invertible $x \in V$, we have that $y \mapsto \phi(x,y)$ is a polynomial map of degree $d-1$. We put

$$\phi(x,y) = \sum_{i=0}^{d-1} \phi_i(x,y),$$

where $y \mapsto \phi_i(x,y)$ is a homogeneous polynomial map $V \to V$ of degree i. ϕ_i is a rational map $V \times V \to V$.

3.2 Lemma. *If $y \in V$ is fixed, $x \mapsto \phi_i(x,y)$ is a homogeneous rational map of degree $i+1$ $(0 \leq i \leq d-1)$.*

Let $t \in K^*$. We have by the homogeneity of j, n and N

$$\phi(t\,x, y) = N(t\,x)\,n(j(t\,x) + y) = t^d N(x)\,n(t^{-1} \cdot jx + y)$$
$$= t\,N(x)\,n(jx + t\,y) = t\,\phi(x, t\,y).$$

This implies the assertion.

3.3 Lemma. *If $x, y \in V$ are invertible we have*

(3) $\qquad N(x)\, n(j\, x+y) + N(y)\, P(x)\, n(x+j\, y) = N(x)\, N(j\, x+y)\, x.$

Let $\Psi(x, y)$ denote the left-hand or right-hand side of (3). Using (2), 1.5, 1.9 and 1.16(i) one sees that for $g \in G$ we have

$$\Psi(g\, x, g'\, y) = g\, \Psi(x, y).$$

By axiom (J3) it follows from this formula that if (3) holds for $x = e$, it holds for all invertible x in a suitable nonempty open subset of V, hence by continuity for all invertible x. So it suffices to prove that (3) is true for $x = e$, i.e. that

$$n(e+y) + N(y)\, n(e+j\, y) = N(e+y)\, e.$$

By 1.7 and the definition of n and N this is a consequence of axiom (J2).

3.4 Theorem. *The rational map $P: V \to \mathrm{End}(V)$ is a quadratic polynomial map.*

Let $y \in V$, let $x \in V$ be invertible. As in 1.11, put

$$N(x)\, N(j\, x + y) = \sum_{i=0}^{d} \Phi_i(x, y),$$

where Φ_i is a polynomial function on $V \times V$, homogeneous of degree i in either variable. It follows from (3) that

(4) $\qquad \begin{cases} \phi_0(x, y) = x, \\ \phi_i(x, y) + P(x)\, \phi_{i-1}(y, x) = \Phi_i(x, y)\, x \quad (1 \leq i \leq d). \end{cases}$

Taking $i = 1$ in (4) we obtain

(5) $\qquad P(x)\, y = \sigma(x, y)\, x - \phi_1(x, y),$

where $\sigma = \Phi_1$ is the standard symmetric bilinear form.

Taking $i = 2$ in (4) we find, if y is invertible

$$\phi_1(x, y) = P(x)^{-1} \{\Phi_2(y, x)\, y - \phi_2(y, x)\}.$$

Now Φ_i is a polynomial function and $x \mapsto \phi_i(y, x)$ is a polynomial map. It then follows from the last formula that for fixed invertible $y \in V$, we have that $x \mapsto \phi_1(x, y)$ is a polynomial map. Hence the denominator of the rational map $\phi_1: V \times V \to V$ is independent of the first variable. By what was said in 3.1, this denominator involves only the first variable, hence it must be a constant. Consequently ϕ_1 is a polynomial map. That P is a homogeneous quadratic map now follows from (5), using 3.2.

§3. The Quadratic Map of a J-structure

3.5 An irreducibility criterion. Let $G°$ be the identity component of the structure group G of $\mathscr{S}$, let $G_1 \subset G°$ be the inner structure group, see 1.19. The next result makes use of (5). Its corollary 3.7 can also be viewed as a simplicity criterion for J-structures. We discuss the relation between simplicity and irreducible action of $G°$ in §12.

3.6 Theorem. *Suppose that the norm N is an irreducible polynomial function. Let $W \subset V$ be a proper subspace which is G_1-stable. Then $\sigma(V, W) = 0$.*

We shall prove the following assertion:

(∗) if $a, b \in V$ and $\sigma(P(x)a, b) = 0$ for all $x \in V$, then either $\sigma(V, a) = 0$ or $\sigma(V, b) = 0$.

We first show that (∗) implies 3.6. Let W be a proper G_1-stable subspace. Assume that $\sigma(V, W) \neq 0$. Then there exists $b \in V$ such that $\sigma(V, b) \neq 0$ and $\sigma(W, b) = 0$. If $a \in W$, then $P(x)a \in W$ for all invertible $x \in V$ (since W is G_1-stable), whence $P(x)a \in W$ for all $x \in V$. It follows that $\sigma(P(x)a, b) = 0$, whence $\sigma(V, a) = 0$ by (∗). This implies that $\sigma(V, W) = 0$, a contradiction.

To prove (∗), we observe that (5) implies that, under the assumptions of (∗),
$$\sigma(x, a)\,\sigma(x, b) = \sigma(\phi_1(x, a), b).$$

By the definition of σ and ϕ_1 we have
$$\sigma(x, y) = N(x)(dN)_{jx}(y),$$
$$\phi_1(x, y) = N(x)(dn)_{jx}(y).$$

Replacing x by jx and using 1.8 we see that
$$(dN)_x(a) \cdot (dN)_x(b) = N(x)\,\sigma((dn)_x(a), b).$$

It follows that the irreducible function N divides the product of two polynomial functions of strictly smaller degree. Consequently, one of these is 0. Hence $(dN)_x(a) = 0$ or $(dN)_x(b) = 0$, which is equivalent to $\sigma(V, a) = 0$ or $\sigma(V, b) = 0$. This establishes (∗) and concludes the proof of 3.6.

3.7 Corollary. *Suppose that N is an irreducible polynomial function and that σ is nondegenerate. The G_1 and $G°$ act irreducibly in V.*

That G_1 acts irreducibly under these assumptions is a direct consequence of 3.6. Since G_1 is a subgroup of $G°$, the same holds for the latter group (of course, G itself acts irreducibly).

3.8 Examples. Consider the J-structures $\mathscr{M}_r, \mathscr{S}_r, \mathscr{A}_r, \mathscr{O}_{2,r}, \mathscr{O}'_{2,r}$, introduced in §2. Using what was established there, one finds that G_1 acts irreducibly

in the following cases:
$\mathcal{M}_r$ and $\mathcal{A}_r$ for $r \geq 1$;
$\mathcal{S}_r$ for $r \geq 1$ and $\operatorname{char}(K) \neq 2$;
$\mathcal{O}_{2,r}$ for $r \geq 3$;
$\mathcal{O}'_{2,r}$ for $r \geq 2$.

Next consider $\mathcal{S}_r$ for $r \geq 2$ and $\operatorname{char} K = 2$. The standard symmetric bilinear form σ is then degenerate, as one sees from the formula given in 2.10. It can be checked, using 3.6 that the only nontrivial subspace of the space $\mathbb{S}_r$ of symmetric matrices which is G_1-stable is the space $\mathbb{A}_r$ of alternating matrices.

Similarly, one can discuss the case of $\mathcal{O}''_{2,r}$ ($r \geq 1$, $\operatorname{char} K = 2$). Then the only nontrivial G_1-stable subspace is the 1-dimensional space $K(e_r + e_{2r})$ (with the notations of 2.20).

3.9 Some formulas. We call the map P of 3.4 P the *quadratic map* of the J-structure $\mathcal{S}$. We write

$$P(x, y) = P(x+y) - P(x) - P(y) \quad (x, y \in V).$$

Since P is quadratic, we have that $(x, y) \mapsto P(x, y)$ is a symmetric bilinear map $V \times V \to \operatorname{End}(V)$. By the definition of differentials (see 0.9) we have

$$P(x, y) = (dP)_x(y).$$

Notice that
$$P(x, x) = 2 P(x).$$

We collect now a number of identities involving P. First observe that by 1.16(vi) and 3.4 we have

(6) $\qquad P(P(x) y) = P(x) P(y) P(x) \quad (x, y \in V).$

(6) is an identity of degree 4 in x. Polarization, i.e. differentiation with respect to x, leads to a number of consequences, of which we only mention the following identity

(7) $\quad \begin{aligned} & P(P(x, y) z) + P(P(x) z, P(y) z) \\ & = P(x) P(z) P(y) + P(y) P(z) P(x) + P(x, y) P(z) P(x, y). \end{aligned}$

For the next set of formulas we start with the formula of 1.16(iii) viz.

$$P(x) j x = x,$$

if x is invertible. We find by differentiation that

$$P(x, y) j x + P(x) (dj)_x(y) = y,$$

whence, by the definition of $P(x)$,

(8) $\qquad P(x, y) j x = 2 y \quad (x \in V \text{ invertible}, y \in V).$

§3. The Quadratic Map of a J-structure

Differentiation of (8) gives in the same way

(9) $\quad P(y,z)jx = P(x,y)P(x)^{-1} \cdot z \quad (x \in V \text{ invertible}, y, z \in V).$

Taking $x=e$ in (9) one obtains, using 1.16(iv)

(10) $\quad\quad\quad\quad P(y,z)e = P(e,y)z = P(e,z)y.$

Putting $y=e$ in (9) and using (10) we obtain

$$P(jx, e) = P(x, e) P(x)^{-1} \quad (x \in V \text{ invertible}).$$

Using 1.16(i) and axiom (J3) it follows by an argument using G, similar to that used in the proof of 3.3, that

(11) $\quad\quad\quad\quad P(x) P(jx, jy) P(y) = P(x, y),$

whence

(12) $\quad\quad\quad\quad P(x) P(jx+jy) P(y) = P(x+y),$

if x and y are invertible.

Taking successively $x=e$ and $y=e$ in (11) one sees that

(13) $\quad P(x) P(jx, e) = P(jx, e) P(x) = P(x, e) \quad (x \text{ invertible}),$

from which it follows that $P(x)$ and $P(x, e)$ commute for all $x \in V$.

Next it follows from 1.16(v) and 1.5 that there is a rational function c on V such that

$$N(P(x)y) = c(x) N(y) \quad (x \text{ invertible}, y \in V).$$

Taking $y=jx$, it follows from 1.16(iii) and 1.8 that $c(x) = N(x)^2$. Hence

(14) $\quad\quad\quad\quad N(P(x)y) = N(x)^2 N(y) \quad (x, y \in V).$

Finally we mention the following formula, a variant of which occurred in 1.27,

(15) $\quad\quad\quad\quad j(x+P(x)y) + j(x+jy) = jx$

(valid if $x, y, x+jy, x+P(x)y$ are invertible). For $x=e$ this is §1, (3) (by 1.16(iv)). The general case follows by using 1.16(i) and axiom (J3).

(15) can be rewritten as

(16) $\quad\quad\quad\quad P(x) y = x - j(jx + j(jy - x))$

(valid if the appropriate elements are invertible). We call (16) *Hua's formula*. If $\mathscr{S} = \mathscr{J}(A)$, the J-structure defined by the associative algebra A (see 2.1), it follows from 2.2 that (16) reduces to Hua's identity

$$xyx = x - (x^{-1} + (y^{-1} - x)^{-1})^{-1}.$$

3.10 Powers of an element. Let X be an indeterminate over K. By 0.9 we have a formal power series development.

(17) $$j(e-xX) = \sum_{m=0}^{\infty} x^m X^m,$$

where $x \mapsto x^m$ is a rational map $V \to V$.

3.11 Proposition. (i) $x \mapsto x^m$ *is a homogeneous polynomial map of degree* m, *we have* $x^0 = e$, $x^1 = x$;
(ii) *The following relations hold for* $x \in V$, $i, j, m \geq 0$

(18) $$P(x) x^m = x^{m+2},$$

(19) $$P(x^m) = P(x)^m,$$

(20) $$P(x^i, x^j) x^m = 2 x^{i+j+m};$$

(iii) *Any two of the linear transformations* $P(x)$, $P(x^i, x^j)$ ($i, j \geq 0$) *commute.*

By 1.16(iii) we have for $x \in U$

$$P(e-xX) \cdot \sum_{m=0}^{\infty} x^m X^m = e - xX.$$

Equating coefficients of powers of X we find, since $P(e) = \mathrm{id}$,

(21) $$\begin{cases} x^0 = e, \\ x^1 - P(e, x) e = -x, \\ x^m - P(e, x) x^{m-1} + P(x) x^{m-2} = 0 \quad (m \geq 2). \end{cases}$$

Using (10) we conclude that $x^1 = x$. By induction on m it then follows from (21) and 3.4 that $x \mapsto x^m$ is a homogeneous polynomial map of degree m. This proves (i).

From (8), with $e - xX$ instead of x, we obtain

(22) $$P(x, y) x^{m-1} = P(e, y) x^m \quad (m \geq 1).$$

Taking $y = e$, this gives

(23) $$P(e, x) x^{m-1} = P(e, e) x^m = 2 x^m \quad (m \geq 1).$$

Substituting this into the last formula of (21) we obtain (18). (19) is a consequence of (6) and (18) (using that $P(e) = \mathrm{id}$).

It follows from (6) that

(24) $$P(x^{i+2} x^{j+2}) = P(x) P(x^i, x^j) P(x).$$

§3. The Quadratic Map of a J-structure

This implies, using (18), that in order to prove (20), it suffices to consider the cases $i=0$ and $i=1$. From (22) we conclude that

$$P(x, x^j) x^m = P(e, x^j) x^{m+1},$$

so that finally it suffices to prove (20) for $i=0$. But by (10) we have

$$P(e, x^j) x^m = P(x^j, x^m) e.$$

Using again (6), we can reduce the proof of (20) to the case where $\text{Min}(j, m) = 0$ or 1. In that case the result follows from (10) and (23).

To prove (iii) we replace in (11) x by $e - xX$ and y by $e - xY$, where X and Y are indeterminates over K. Comparison of coefficients of the monomials $X^m Y^n$ then shows that $P(x^m, x^n)$ is a linear combination of products of $P(x)$ and $P(x, e)$. It then suffices to prove that $P(x)$ and $P(x, e)$ commute, which follows from (13), as we already observed. This proves 3.11.

3.12. If $x \in V$, we call x^m the m-th *power of* x. x^2 is the *square* of x. If $\mathscr{S} = \mathscr{J}(A)$, the J-structure defined by the associative algebra A, then this notion of power of an element coincides with the usual one, as follows from (18), using the explicit description of P in that case (see 2.2).

We next introduce negative powers. If $x \in V$ is invertible we put

$$x^{-1} = jx, \quad x^{-m} = (jx)^m \quad (m \geq 0).$$

It follows from (18), using $P(jx) = P(x)^{-1}$ (see 1.15) that

$$P(x) x^{-m} = x^{-(m-2)}.$$

One then proves by induction on m that

$$P(x)^m x^{-m} = x^m,$$

whence by (19) and 1.16 (iii),

$$x^{-m} = j(x^m) = (x^m)^{-1}.$$

We see that (18) holds for all $m \in \mathbb{Z}$ if x is invertible. The same is then true for (19). It is also easily seen that (20) remains true for $i, j, m \in \mathbb{Z}$, if x is invertible.

3.13 Idempotent and nilpotent elements. We say that $x \in V$ is *idempotent* if $x^2 = x$. It then follows from (19) that $P(x^m) = P(x)$ for all $m \geq 2$. Moreover (18) then implies that we have for $m \geq 1$

$$x^{2m} = P(x)^m e = P(x) e = x^2 = x,$$
$$x^{2m+1} = P(x)^m x = P(x)^m x^2 = x^{2m+2} = x.$$

Hence $x^m = x$ for all $m \geq 1$.

3.14 Proposition. $x \in V$ *is idempotent if and only if the following holds: for all* $t, u \in K^*$ *the element* $tx + u(e-x)$ *of* V *is invertible and* $j(tx + u(e-x)) = t^{-1} \cdot x + u^{-1}(e-x)$.

Let x be idempotent. Since $x^m = x$ for all $m \geq 1$, we conclude from (17) that, X denoting an indeterminate over K, we have

$$j(e - xX) = e + x \sum_{m=1}^{\infty} X^m = e + x(1-X)^{-1} X.$$

It follows that the rational map $K \to V$ defined by $t \mapsto j(e - tx)$ is regular for all t except $t = 1$. Then $(t, u) \mapsto j(tx + u(e-x))$ defines a rational map $K \times K \to V$ which is regular if $tu \neq 0$ and

$$j(tx + u(e-x)) = t^{-1} x + u^{-1}(e-x) \quad (t, u \in K^*).$$

This proves one half of the assertion.

Let x have the property of 3.14. Then $t \mapsto j(e - tx)$ is a rational map $K \to V$ which is regular except for $t = 1$, and

$$j(e - tx) = e + x(1-t)^{-1} t.$$

It follows that

$$j(e - xX) = e + x \sum_{m=1}^{\infty} X^m,$$

whence by (17)

$$x^m = x \quad (m \geq 1),$$

which shows that x is idempotent.

We say that $x \in V$ is *nilpotent* if $x^m = 0$ for some $m \geq 1$.

3.15 Proposition. *The following properties of $x \in V$ are equivalent:*
(i) *x is nilpotent;*
(ii) *for all $t \in K$, the element $e - tx$ is invertible;*
(iii) *for all $t \in K$ we have $N(e + tx) = 1$.*

Suppose that $x^m = 0$. Then it follows from (18) that

$$x^{i+2m} = P(x^m) x^i = 0 \quad (i \geq 0).$$

Hence the formal power series in the right-hand side of (17) is now a polynomial. This implies that

$$t \mapsto j(e - tx)$$

defines a polynomial map of K into V, which proves (ii).

Conversely, if (ii) holds, then the rational map $t \mapsto j(e - tx)$ must be a polynomial map. Hence the formal power series in the right-hand side of (17) is a polynomial, so that $x^m = 0$ for sufficiently large m. This shows that (ii) implies (i).

§3. The Quadratic Map of a J-structure

Next observe that (1.8) implies

(25) $$N(e+tx)N(j(e+tx))=1.$$

Now if (ii) holds, it follows from what was said before that

$$t \mapsto N(j(e-tx))$$

is a polynomial function on K. But then (25) implies that $N(e+tx)$ must be a constant independent of t. Putting $t=0$ we see that (iii) holds. Hence (ii) implies (iii).

Finally, if (iii) holds, then the definition of N (see 1.4) shows that all elements $e-tx$ are invertible, so that (ii) holds.

3.16 Corollary. *Let τ be the trace of $\mathscr{S}$. If $x \in V$ is nilpotent we have $\tau(x)=0$.*

This follows from 3.15(iii), because $\tau(x)=(dN)_e(x)$ (see (11)).

Notes

3.4 generalizes a well-known property in Jordan algebras, due to Koecher (see [8, p. 90]). The formulas discussed in 3.9 are familiar ones from Jordan algebra theory, see e.g. [loc. cit., IV].

The definition of powers given in 3.10 is a natural one in the context of J-structures. The characterization of idempotents given in 3.14 is basic for our treatment of the Peirce decomposition (in §10) and of the classification of J-structures (in §11 and §12).

§4. The Lie Algebras Associated with a J-structure

The results of this section will not be used in an essential manner until §14.

Let $\mathscr{S} = (V, j, e)$ be a J-structure with structure group G. We use the notations of §1 and §3.

If H is a linear algebraic group we denote by $L(H)$ its Lie algebra, see [1, p. 118]. $L(G)$ is identified with a subalgebra of $\mathfrak{gl}(V)$ (notation of 0.2).

4.1. Let $\mathfrak{g}$ be the subspace of $\text{End}(V)$ consisting of the $X \in \text{End}(V)$ such that there exists $Y \in \text{End}(V)$ with

(1) $$X(jx) = (dj)_x(Yx),$$

for all invertible $x \in V$. We shall show that $\mathfrak{g}$ is a Lie algebra. But before doing so we have to introduce some notations.

Let N be the norm of $\mathscr{S}$, let $jx = N(x)^{-1} \cdot n(x)$. Let d be the degree of N. Denote by A the set of all homogeneous polynomial maps $a: V \to V$ of degree $d-1$ and by B the set of all homogeneous polynomial functions b on V of degree d. A and B are vector spaces. Let P be the projective space defined by $A \times B$, let p_0 be the canonical image of (n, N) in P.

Put $M = GL(V) \times GL(V)$. M acts in $A \times B$ by

$$((g_1, g_2), (a, b)) \mapsto (g_1 \circ a \circ g_2^{-1}, b \circ g_2^{-1}).$$

This induces an action of M in P, denoted by $(h, p) \mapsto h \cdot p$. The structure group G of $\mathscr{S}$ is then isomorphic to the isotropy group of p_0 in M. Let λ be the morphism of M into P defined by $\lambda h = h \cdot p_0$. The differential $(d\lambda)_1$ of λ at the neutral element 1 is a linear map of the Lie algebra $L(M)$ into the tangent space $T(P)_{p_0}$ of P at p_0. We identify $L(M)$ with $\mathfrak{gl}(V) \times \mathfrak{gl}(V)$.

4.2 Lemma. (i) $\mathfrak{m} = \text{Ker}(d\lambda)_1$ consists of all $(X, Y) \in \mathfrak{gl}(V) \times \mathfrak{gl}(V)$ such that (1) holds;
(ii) $\mathfrak{m}$ is a Lie algebra;

§4. The Lie Algebras Associated with a J-structure

(iii) *The projection of* $\mathfrak{gl}(V) \times \mathfrak{gl}(V)$ *onto its first factor induces an isomorphism of* $\mathfrak{m}$ *onto* $\mathfrak{g}$, *hence* $\mathfrak{g}$ *is a Lie algebra;*

(iv) $L(G) \subset \mathfrak{g}$.

Let $K[\delta]$ be the K-algebra of dual numbers ($\delta^2 = 0$). Let $M(K[\delta])$ denote the group of $K[\delta]$-points of M. If $X_1 \in L(M)$ then $\mathrm{id} + \delta X_1 \in M(K[\delta])$, where id is the identity map of V (see [1, p. 127]). Now $X_1 \in \mathfrak{m}$ if and only if $\mathrm{id} + \delta X_1$ fixes $p_0 \in P$. Let $X_1 = (X, Y)$.

The definition of the action of M shows that $X_1 \in \mathfrak{m}$ if and only if

$$(\mathrm{id} + \delta X) j x = j(x + \delta Y x).$$

Let T be an indeterminate over K. It follows that the formal power series for $j(x + (Yx)T)$ (see 0.9) starts off with $jx + X(jx)T$, which shows that (1) holds. This implies (i).

To prove (ii) we use two copies $K[\delta]$ and $K[\delta']$ of the dual numbers. These we imbed in the standard way in $K[\delta] \otimes K[\delta']$. If $X_1, Y_1 \in L(M)$, we have in $M(K[\delta] \otimes K[\delta'])$ that

$$(\mathrm{id} + \delta X_1)(\mathrm{id} + \delta' Y_1)(\mathrm{id} + \delta X_1)^{-1}(\mathrm{id} + \delta' Y_1)^{-1} = \mathrm{id} + (\delta \otimes \delta')[X_1, Y_1].$$

Hence if $X_1, Y_1 \in \mathfrak{m}$, we have that the element $\mathrm{id} + (\delta \otimes \delta')[X_1, Y_1]$ of $M(K[\delta \otimes \delta'])$ fixes p_0. Since $\delta \mapsto \delta \otimes \delta'$ defines an isomorphism of $K[\delta]$ into $K[\delta \otimes \delta']$, it follows that $\mathrm{id} + \delta[X_1, Y_1] \in G(K[\delta])$, whence $[X_1, Y_1] \in \mathfrak{m}$. This proves (ii).

(iii) is clear and (iv) follows from the connection between G and the isotopy group of p_0 in M.

We call $\mathfrak{g}$ the *structure algebra* of $\mathscr{S}$. Remark that by §1,(16) the defining formula (1) can be rewritten as

(2) $$P(x)(X(jx)) = Yx.$$

Let P be the quadratic map of $\mathscr{S}$. Put

$$L(x, y) = P(x, jy) P(y).$$

($x \in V$, y invertible).

4.3 Lemma. (i) *If* $g \in G$ *then*

(3) $$L(gx, g'y) = g \circ L(x, y) \circ g^{-1}$$

($g \mapsto g'$ *is the standard automorphism of* G);

(ii) $L(\ ,\ )$ *can be extended to a bilinear map of* $V \times V$ *into* $\mathrm{End}(V)$ *satisfying*

(4) $$L(x, y) z = L(z, y) x = P(x, z) y.$$

(3) follows from the definition of L, using 1.16. That (4) holds with $y = e$ is a consequence of §3, (10). By (3) and axiom (J 3) we then obtain (4) for y in a nonempty open subset of V, for all $x, z \in V$. This implies (ii).

Let $\mathfrak{g}_1$ be the subspace of End(V) spanned by all $L(x, y)$. Let G_1 be the inner structure group of $\mathscr{S}$ (see 1.19). G acts on $L(G)$ by the adjoint representation Ad. Identifying G and $L(G)$ with subsets of End(V), we have
$$\mathrm{Ad}(g) X = g \circ X \circ g^{-1} \quad (g \in G, X \in L(G)).$$

4.4 Proposition. (i) $\mathfrak{g}_1 \subset L(G_1) \subset L(G) \subset \mathfrak{g}$;
(ii) $\mathfrak{g}_1$ *is stable for the adjoint action of G, and $\mathfrak{g}_1$ is an ideal in $L(G)$*;
(iii) *If $g \in G_1$, $X \in \mathfrak{g}$ then $\mathrm{Ad}(g) X - X \in \mathfrak{g}_1$ and $\mathfrak{g}_1$ is an ideal in $\mathfrak{g}$.*

P defines a morphism of a neighbourhood of e into G_1, whose differential at e is $x \mapsto P(x, e)$ (the tangent space at e being identified with V). It follows that $L(x, e) \subset L(G_1)$ for all $x \in V$. Since G_1 is a normal subgroup of G (see 1.20) we have by (3) that
$$g \circ L(x, e) \circ g^{-1} = L(g x, g' \cdot e) \in L(G_1),$$
whence, using axiom (J 3), $L(x, y) \in L(G_1)$ for all $x, y \in V$. This proves that $\mathfrak{g}_1 \subset L(G_1)$. The other inclusions of (i) being clear, we have established (i).

The first assertion of (ii) has also been proved. The second one is a consequence (compute the differential of the morphism $g \mapsto \mathrm{Ad}(g) X$ of G_1 into $\mathfrak{g}_1$, for fixed $X \in \mathfrak{g}_1$; see [1, 3.9(1), p. 134–135]).

Let $X \in \mathfrak{g}$, let $Y \in \mathrm{End}(V)$ be such that (2) holds. Differentiation of (2) shows that we have

(5) $$(P(x) \circ X \circ P(x)^{-1}) y = Yy + P(x, y)(Xjx).$$

Now by (2) and (4) we have
$$P(x, y)(Xjx) = -P(x, y) P(x)^{-1} Yx = -L(y, jx) Yx = -L(Yx, jx) y.$$

Moreover, (5) with $x = e$ implies that

(6) $$Yy = X y - P(e, X e) y.$$

(5) can now be rewritten as
$$(P(x) \circ X \circ P(x)^{-1}) y = X y - L(e, X e) y - L(Yx, jx) y,$$
which implies the first assertion of (iii). The second one is then a consequence. As in the proof of (ii) one also sees that we have $[L(G_1), \mathfrak{g}] \subset \mathfrak{g}_1$. It follows, in particular, that $\mathfrak{g}_1$ is a Lie algebra. We call it the *inner structure algebra* of $\mathscr{S}$.

4.5 Lemma. *Let $X \in \mathfrak{g}$. Then*
$$X x^2 = P(e, x) X x - P(x) X e.$$

This readily follows from (2), using §3, (17).

We next discuss some Lie algebras related to the automorphism group of $\mathscr{S}$. Let H be that group. Clearly $H \subset G$.

§4. The Lie Algebras Associated with a J-structure

4.6 Proposition. *H consists of the elements of G that fix e.*

Let $g \in G$, $g\,e = e$. We have to prove that g is an automorphism. From §1, (3) and the definition of the standard automorphism of G we see that

$$j(e+g\,x)+j(e+j(g'\cdot x))=g'\cdot e.$$

Using again §1, (3) it follows that

(7) $$j(e+g\,x)-j(e+g'\cdot x)=g'\cdot e-e.$$

Put $h=g'\cdot g^{-1}$. The last formula implies that

$$j\,x=j(e-h\,e+h\,x)+g'\cdot e-e$$

(this is to be considered as an identity of rational functions).

The left-hand side being homogeneous of degree -1, it follows that

$$j(t(e-h\,e)+h\,x)-j(e-h\,e+h\,x)-g'\cdot e+e=t^{-1}(g'\cdot e-e),$$

if $t \in K^*$. For suitable $x \in V$ the left-hand side is regular in $t=0$. This can only be if $g'\cdot e=e$. It then follows from (7) that $g'=g$, which means that g is an automorphism.

We put $H_1 = H \cap G_1$ and

$$\mathfrak{h}=\{X\in\mathfrak{g}\,|\,X\,e=0\},$$
$$\mathfrak{h}_1=\mathfrak{g}_1\cap\mathfrak{h}.$$

Clearly $\mathfrak{h}$ and $\mathfrak{h}_1$ are Lie algebras. We have $L(H) \subset \mathfrak{h}$.

4.7 Lemma. (i) *$\mathfrak{h}$ consists of the linear transformations X of V such that $X\,e=0$ and that*

(8) $$P(x)\,X(j\,x)=-X\,x,$$

for all invertible $x\in V$;
(ii) *If $\mathrm{char}(K)\neq 2$ then (8) implies that $X\,e=0$.*

Assume that $X \in \mathfrak{g}$, $X\,e=0$. Let Y be such that (2) holds. The formula (6), established in the proof of 4.4, shows that $Y=X$, whence (8).

If (8) holds, then (6) gives that $P(e, X\,e)\,y=0$. In particular, $2(X\,e)=P(e, X\,e)\,e=0$, whence (ii).

4.8 Lemma. *If $h \in H_1$, $X \in \mathfrak{h}$ then $\mathrm{Ad}(h)\,X-X \in \mathfrak{h}_1 \cdot \mathfrak{h}_1$ is an ideal in $\mathfrak{h}$.*

This follows from 4.4(iii).
We call the elements of $\mathfrak{h}$ *derivations* of $\mathcal{S}$ and those of $\mathfrak{h}_1$ *inner derivations*. $\mathfrak{h}$ is the *derivation algebra* of $\mathcal{S}$.

4.9 Fields of definition for G and H. (The results which follow will not be needed until §15.)

The Lie algebras $\mathfrak{g}$ and $\mathfrak{h}$ are needed to establish results about fields of definition for G and H. We say that G (or H) is *smooth* if $\mathfrak{g} = L(G)$ (or $\mathfrak{h} = L(H)$).

Let $k \subset K$ be a subfield such that $\mathscr{S}$ is defined over k. We denote by k_s a separable closure of k in K.

4.10 Theorem. *If G is smooth then G is defined over k.*

We use the notations of 4.1. We can then view G as the isotropy group of the point p_0, in the action of M on P. The assertion then follows by application of [1, 6.7, p. 180], using 4.2(i) and 4.2(iii).

Next let $\mathscr{S}' = (V', j', e')$ be another J-structure, which is also defined over k and which is isomorphic to $\mathscr{S}$ over K. Let X be the set of isomorphisms of $\mathscr{S}$ onto $\mathscr{S}'$, this is a subset of the vector space $\mathrm{Hom}(V, V')$ of K-homomorphisms of V into V'.

4.11 Theorem. *X is an algebraic subvariety of $\mathrm{Hom}(V, V')$. If H is smooth then X is defined over k.*

Use again the notations of 4.1. Let A', B', P' be for $\mathscr{S}'$ as A, B, P for $\mathscr{S}$. Let $S \subset \mathrm{Hom}(V, V')$ be the set of isomorphisms g of V onto V' with $g e = e'$. S is an algebraic subvariety of $\mathrm{Hom}(V, V')$ which is defined over k. Let π be the morphism of $S \times P$ into P' defined by

$$\pi(g, [a, b]) = [g \circ a \circ g^{-1}, b \circ g^{-1}]$$

(where $[a, b]$ denote the element of P defined by $(a, b) \in A \times B$, similarly for P'). j and j' are identified with elements p_0, p'_0 of P and P'. We then have

$$X = \{g \in S \mid \pi(g, p_0) = p'_0\}.$$

This shows that X is an algebraic subvariety of S, hence of $\mathrm{Hom}(V, V')$. Let ρ be the morphism of S into P' defined by $\rho(g) = \pi(g j)$. It is clear that $\dim \rho(S) = \dim S - \dim H$. Now by a computation as the one used to establish 4.2(i) one proves that the kernel of $(d\rho)_x$ is isomorphic to $\mathfrak{h}$ as a vector space, in any point $x \in X$. It then follows from the smoothness of H that $(d\rho)_x$ is surjective. As in [1, p. 181] it then follows that X is defined over k.

4.12 Corollary. *If H is smooth then H is defined over k.*

This is the case $S' = S$ of 4.10.

4.13 Corollary. *Assumptions of 4.11. There is an isomorphism of $\mathscr{S}$ onto $\mathscr{S}'$ which is defined over k_s.*

X is defined over k by 4.11, hence $X(k_s) \neq \phi$ (in fact, $X(k_s)$ is Zariski dense in X, see [1, 13.3, p. 52]).

§4. The Lie Algebras Associated with a J-structure

4.14. With the previous notations, assume now moreover that k is a perfect field. Then G and H are always defined over k, as follows from [1, p. 47] (observe that G and H are clearly k-closed). So for perfect base fields, the smoothness condition can be omitted in 4.10 and 4.12. The same is true for 4.11.

Notes

It can be shown that (2) is equivalent to

$$P(Xx, x) = X \circ P(x) - P(x) \circ Y \quad (x \in V).$$

With this formula one recovers the definition of the structure algebra of a Jordan algebra, given in [14, p. 437].

The use of the Lie algebra in 4.10 and 4.11, to deal with problems about (non-perfect) fields of definition is a familiar principle from the theory of linear algebraic groups, see e.g. [3].

§5. J-structures of Low Degree

We keep the notations of Section 4. In this section we shall discuss J-structures of degree $d \leq 3$, the most interesting case being $d=3$. The trivial case $d=1$ has already been discussed in 2.19.

We assume that $\dim V > 1$. We also suppose that the norm N is a *nondegenerate* polynomial function (see 0.15). We shall see in §9 that this means that the J-structure $\mathscr{S}$ is semisimple.

5.1. We first assume that $d=2$. In that case the norm N is a quadratic form on V. We put

$$N(x, y) = N(x+y) - N(x) - N(y) \quad (x, y \in V),$$

then $(x, y) \mapsto N(x, y)$ is a symmetric bilinear form on $V \times V$. Let

$$N_x(y) = N(x, y),$$

then N_x is a linear function on V.

Since $d=2$, the numerator n of j is a *linear* transformation of V. From (J 1) and 1.8 one obtains that $n^2 = \mathrm{id}$. From §3, (3) with $x = e$ one then finds that

$$n(x) = -x + N(x, e) e,$$

hence j has the same form as in 2.17. One then shows that, as in 2.18, the structure group G of j consists of the linear transformations $g \in GL(V)$ such that there exists $a \in K^*$ with

$$N(g x) = a N(x) \quad (x \in V).$$

Put

$$R = \{x \in V | N_x = 0\}.$$

Then R is a G-stable subspace of V. If $N_x \neq 0$ for some $x \in V$ we have $R \neq V$ and then axiom (J 3) implies that $e \notin R$. Hence $N_e \neq 0$ and we are in the situation of 2.17. If $N_x = 0$ for all $x \in V$, then $2 = N(e, e) = 0$, so that $\mathrm{char}(K) = 2$. Then N is the square of a linear function. The nondegeneracy of N then implies that $\dim V = 1$, which is impossible. It follows that a J-structure of degree 2 with nondegenerate norm is isomorphic to some $\mathcal{O}_r, \mathcal{O}'_{2,r}, \mathcal{O}''_{2,r}$ (the latter two if $\mathrm{char}(K) = 2$).

§5. J-structures of Low Degree

5.2. From now on we assume in this section that $d=3$. Hence N is a cubic form on V. We put

$$N(x, y) = (dN)_x(y) \quad (x, y \in V).$$

Then $(x, y) \mapsto N(x, y)$ is a polynomial function on $V \times V$, which is homogeneous of degree 2 in its first variable and linear in the second one. We have for $x, y \in V$

Put
$$N(x+y) = N(x) + N(x, y) + N(y, x) + N(y).$$

$$N(x, y, z) = N(x+y, z) - N(x, z) - N(y, z) \quad (x, y, z \in V),$$

then $(x, y, z) \mapsto N(x, y, z)$ is a trilinear form on $V \times V \times V$.

From the (easily checked) identity

$$N(x, y, z) = N(x+y+z) - N(y+z) - N(x+z) - N(x+y)$$
$$+ N(x) + N(y) + N(z)$$

one concludes that the trilinear form is symmetric in its three variables. We have

$$N(x, x, y) = 2N(x, y), \quad N(x, x) = 3N(x), \quad N(x, x, x) = 6N(x).$$

Define a symmetric bilinear form σ on $V \times V$ by

(1) $$\sigma(x, y) = N(e, x) N(e, y) - N(e, x, y),$$

by 1.18 this is the standard symmetric bilinear form σ, defined in 1.11.

The numerator n of j is now a quadratic map $V \to V$. We define a symmetric bilinear map $(x, y) \mapsto x \times y$ of $V \times V$ into V by

$$x \times y = n(x+y) - n(x) - n(y).$$

By (J1)(ii) we have
(2) $$n(e) = e, \quad N(e) = 1.$$

It follows from (J1)(i) and 1.8 that

(3) $$n(nx) = N(x) x \quad (x \in V).$$

Moreover it follows from the definition of N and σ (see 1.11) that

(4) $$N(x, y) = \sigma(nx, y) \quad (x, y \in V).$$

5.3 Lemma. *Assume that the polynomial function N is reducible. Then $\mathscr{S}$ contains an ideal of codimension 1, hence $\mathscr{S}$ is not simple.*

If N is reducible then one of its irreducible factors must be a linear function l. It follows from 1.12 that we have $l(jx) = l(x)^{-1}$ if x is invertible. The definition of an ideal then shows that $\operatorname{Ker} l$ is an ideal of codimension 1.

Let V be a finite dimensional vector space, let $e \in V$. Suppose that N is a nondegenerate cubic form on V and n a quadratic map $V \to V$. Define a rational map $j: V \to V$ by

$$jx = N(x)^{-1} \cdot n(x) \quad (x \in V).$$

We shall give conditions for (V, j, e) to be a J-structure. We introduce the notations of 5.2.

5.4 Lemma. *Assume that* (2), (3) *and* (4) *hold. Let* $a \in V$.
 (i) *We have* $\sigma(a, V) = 0$ *if and only if* $N(V, a) = 0$;
 (ii) *If* $\mathrm{char}(K) \neq 2, 3$ *then* σ *is nondegenerate;*
 (iii) *If* $\mathrm{char}(K) = 2$ *then* $\sigma(a, V) = 0$ *implies* $N(a) = 0$, *if moreover* $N(a, V) = 0$ *then* $a = 0$;
 (iv) *If* $\mathrm{char}(K) = 3$ *then* $\sigma(a, V) = 0$ *implies* $N(a, V) = 0$, *if moreover* $N(a) = 0$ *then* $a = 0$.

(i) follows from (3) and (4). Assume that $\sigma(a, V) = 0$. From (4) we infer that

(5) $$N(x, y, z) = \sigma(x \times y, z) = \sigma(y \times z, x)$$

whence $N(x, y, a) = 0$ for all $x, y \in V$. In particular we have $2N(a, x) = N(x, x, a) = 0$, also $3N(a) = N(a, a) = 0$. If $\mathrm{char}(K) \neq 2, 3$ this shows that $N(a, V) = 0$, $N(a) = 0$. Hence $N(x + a) = N(x)$ for all $x \in V$ so that $a = 0$ by the nondegeneracy of N. This establishes (ii). The proof of (iii) and (iv) is similar.

The following theorem is the main result of this section.

5.5 Theorem. *Assume that* (2), (3) *and* (4) *hold and that N is irreducible. Then* $\mathscr{S} = (V, j, e)$ *is a J-structure.*

We have to prove the axioms (J1), (J2), (J3), under the assumptions of 5.5. Applying n to both sides of (3) and using (3) with nx instead of x, we find that
$$N(nx) = N(x)^2 \quad (x \in V),$$
whence
$$N(jx) = N(x)^{-1} \quad (\text{if } N(x) \neq 0).$$

(2) then implies (J1)(i). That (J2)(ii) holds follows also from (2).

From (4) we obtain (if $N(x) \neq 0$)

$$N(jx, y) = N(x)^{-1} \sigma(x, y), \quad N(y, jx) = N(x)^{-1} \sigma(nx, ny),$$
whence
$$N(x) N(jx + y) = 1 + \sigma(x, y) + \sigma(nx, ny) + N(x) N(y).$$

It follows that if $N(x) N(y) \neq 0$ we have

(6) $$N(x) N(jx + y) = N(y) N(x + jy).$$

§5. J-structures of Low Degree

Let $N(x) \neq 0$, then j is regular in x. As in §1, (16) define a linear transformation $P(x)$ of V by

$$P(x) = -(dj)_x^{-1} = -(dj)_{jx}.$$

Using the definition of j one finds that

$$-(dj)_x(y) = N(x)^{-2} N(x,y) \, n \, x - N(x)^{-1} \, x \times y.$$

Replacing x by jx and using (3) and (4) one then finds

(7) $$P(x) y = \sigma(x, y) x - n \, x \times y.$$

It follows that P is, in fact, a quadratic map $V \to \mathrm{End}(V)$, as in §3.

Let X be an indeterminate over K. For x and y in a suitable open subset of V we then have a formal power series for $j(x + yX)$, which has the form

$$j(x + yX) = jx - P(x)^{-1} yX + \cdots.$$

Using (6) we obtain that

$$-N(x + yX) N(j(x + yX) - jx) N(x) = N(y) X^3.$$

Comparing coefficients of X^3 we see that

$$N(x)^2 N(P(x)^{-1} y) = N(y).$$

Replacing x by jx this gives

(8) $$N(P(x) y) = N(x)^2 N(y)$$

for $N(x) \neq 0$, hence for all $x, y \in V$.

From the definition of P it follows that $P(x) P(jx) = \mathrm{id}$ (if $N(x) \neq 0$), in particular we have $P(e)^2 = \mathrm{id}$. We next show that $P(e) = \mathrm{id}$. From (1), (4) and (5) we find that

$$\sigma(x, y) = \sigma(x, e) \sigma(y, e) - \sigma(e \times x, y).$$

Using (7) this gives

(9) $$\sigma(P(e) x, y) = \sigma(x, y) \quad (x, y \in V).$$

5.4(ii) shows that $P(e) = \mathrm{id}$ if $\mathrm{char}(K) \neq 2, 3$.

To deal with the remaining cases we use the other statements of 5.4. If $\mathrm{char}(K) = 2$ we have to show, by (iii), that $N(P(e) x - x, y) = 0$. Now

$$N(P(e) x - x, y) = N(P(e) x, y) + N(x, y) - N(P(e) x, x, y).$$

Using (8) and (4) one sees that

$$N(P(e) x, y) = N(x, P(e) y) = N(x, y).$$

Similarly, $N(P(e) x, x, y) = N(x, x, y)$. It follows that

$$N(P(e) x - x, y) = 2N(x, y) - N(x, x, y) = 0.$$

This establishes that $P(e) = \mathrm{id}$ if $\mathrm{char}(K) \neq 2$. If $\mathrm{char}(K) = 3$ the proof is analogous.

We can now establish axiom (J2). (6) with $y = e$ gives

$$N(x) N(e + j x) = N(e + x) \quad (N(x) \neq 0).$$

It follows that in order to prove (J2) we have to establish that

$$n(e + x) + N(x) n(e + j x) = N(e + x) e \quad (\text{if } N(x) \neq 0),$$

which is a direct consequence of (3), (4) and $P(e) = \mathrm{id}$.

It remains to prove (J3). Let G be the structure group of j. We shall establish that $P(x) \in G(N(x) \neq 0)$. This follows from the following auxiliary result.

5.6 Lemma. *If $x, y \in V$ then $P(x) n(P(x) y) = N(x)^2 n y$.*

This is proved by a formal computation. It follows from (3), replacing x by $x + t y$ ($t \in K$), and comparing coefficients of t^2 and t^3 on both sides of the resulting formula, that

(10) $\qquad n x \times n y + n(x \times y) = N(x, y) y + N(y, x) x,$

(11) $\qquad n x \times (x \times y) = N(x) y + N(x, y) x.$

The first formula gives, using (3) and (4), that

$$n(n x \times y) = -N(x)(x \times n y) + N(x) \sigma(x, y) y + \sigma(n x, n y) n x.$$

Similarly, the second one implies that

$$N(x)(x \times (n x \times y)) = N(x)^2 y + N(x) \sigma(x, y) n x,$$

whence

(12) $\qquad x \times (n x \times y) = N(x) y + \sigma(x, y) n x,$

if $N(x) \neq 0$, hence also for all $x \in V$.

By (7) we have

$$n(P(x) y) = \sigma(x, y)^2 n x + n(n x \times y) - \sigma(x, y)(x \times (n x \times y)).$$

Using the previous formulas we obtain

$$n(P(x) y) = -N(x)(x \times n y) + \sigma(n x, n y) n x = P(n x) n y.$$

Since $P(j x) = P(x)^{-1} (N(x) \neq 0)$ the equality of the two extreme terms is equivalent to the formula of the lemma.

§5. J-structures of Low Degree

It follows from 5.6 that
$$P(x)j(P(x)y) = jy$$
which shows that $P(x) \in G$ if $N(x) \neq 0$.

Put $x^2 = P(x)e$ ($x \in V$). In order to establish (J3) it now suffices to prove:
the morphism $f: x \mapsto x^2$ of V into V has a dense image.

Put $P(x, y) = P(x + y) - P(x) - P(y)$. From (7) one obtains that
$$P(x, y)z = \sigma(x, z)y + \sigma(y, z)x - (x \times y) \times z.$$

Using (1) and $P(e) = \mathrm{id}$ one then sees that

(13) $$P(x, e)y = P(x, y)e,$$

whence
$$P(x, e)e = 2x \quad (x \in V).$$

But $P(x, e) = (df)_e(x)$. Hence $(df)_e$ is multiplication by 2. So, if $\mathrm{char}(K) \neq 2$, we have that $(df)_e$ is surjective. It then follows that $f(V)$ is dense in V (see [1, 17.3, p. 75]).

If $\mathrm{char}(K) = 2$ we have to give a more complicated argument. We first derive some properties of P (which have to hold in a J-structure). From axiom (J2), which has already been established, we conclude that
$$(dj)_{e+x} + (dj)_{e+jx} \circ (dj)_x = 0$$
(for x in a suitable open subset of V), whence
$$P(x)P(e + jx) = P(e + x) \quad (N(x) \neq 0)$$
and
$$P(x)P(e, jx) = P(e, x).$$
Differentiating again we deduce
$$P(x, y)P(e, jx) - P(x)P(e, P(x)^{-1}y) = P(e, y) \quad (N(x) \neq 0).$$

From now on assume that $\mathrm{char}(K) = 2$. Replacing y by e and x by jx it follows that
$$P(e, x^2) = P(x)P(e, jx)P(e, x) \quad (\text{if } N(x) \neq 0).$$
Using (13) we obtain
$$P(e, x^2)x = P(x)P(e, jx)P(e, x)x = 2P(x)P(e, jx)P(x)e = 0$$
if $N(x) \neq 0$, hence $P(e, x^2)x = 0$ for all $x \in V$.

By (13) it also follows that
$$P(e, x)e = P(e, x^2)e = P(x, x^2)e = 0,$$

consequently we have for $a, b, c \in K, x \in V$

$$(ae+bx+cx^2)^2 = P(ae+bx+cx^2)e = a^2 e + b^2 x^2 + c^2 (x^2)^2.$$

By (7)
$$x^2 = nx + \sigma(x,e)x + \sigma(e,nx)e,$$
whence
$$(x^2)^2 = (nx)^2 + \sigma(x,e)^2 x^2 + \sigma(e,nx)^2 e.$$

Now by (7)
$$(nx)^2 = P(nx)e = \sigma(nx,e)nx - N(x)(x \times e).$$

Expressing everything in terms of e, x and x^2 we find that there exist polynomial functions α and β such that

$$(ae+bx+cx^2)^2$$
$$= (a^2 + \alpha(x)c^2)e + (N(x) + \sigma(nx,e)\sigma(x,e))c^2 x + (b^2 + \beta(x)c^2)x^2.$$

It follows that if $N(x) \neq \sigma(nx,e)\sigma(x,e)$ there exists $a, b, c \in K$ such that
$$x = (ae+bx+cx^2)^2.$$

By assumption N is irreducible, so the set of $x \in V$ such that $N(x) \neq \sigma(nx,e)\sigma(x,e)$ is a nonempty open subset of V. This proves that $f(V)$ contains a dense open subset of V, if char$(K) = 2$. This concludes the proof of 5.5.

5.7 Corollary. *Under the assumptions of 5.5 the set of squares of elements of V is dense in V.*

This was established in the course of the proof of 5.5.

5.8 Corollary. *Assume moreover that σ is nondegenerate. Then the inner structure group and the identity component $G°$ of the structure group act irreducibly in V.*

Since σ is the standard symmetric bilinear form of $\mathscr{S}$, 5.8 follows from 3.7.

5.9 *Remark.* The assumption that N be irreducible has been used in the proof of 5.5 only in the characteristic 2 case. It would not be difficult to deal with the case that N is reducible.

5.10 Exceptional J-structures. We next discuss an application of 5.5, namely the construction of exceptional J-structures. The method used in the adaptation of the construction of exceptional Jordan algebras by Tits.

§5. J-structures of Low Degree

Let A be a central simple associative algebra over K of degree 3, with identity element 1 (see 0.17). Hence A is isomorphic to the algebra $\mathbb{M}_3$ of 3×3 matrices. Let v and τ denote the reduced norm and reduced trace of A, respectively. There exists a quadratic map adj of A into itself such that for invertible $a \in A$ we have

$$a^{-1} = v(a)^{-1} \cdot \mathrm{adj}\, a$$

(see 2.5).

Put $V = A \oplus A \oplus A$, $e = (1, 0, 0) \in V$. Fix $\alpha \in K^*$. We define a cubic form N on V, a quadratic map $n: V \to V$ and a symmetric bilinear form σ on $V \times V$ as follows. Let $x = (x_0, x_1, x_2)$, $y = (y_0, y_1, y_2)$ be elements of V. Then

(14) $\quad N(x) = v(x_1) + \alpha\, v(x_1) + \alpha^{-1} v(x_2) - \tau(x_0\, x_1\, x_2),$

(15) $\quad n(x) = (\mathrm{adj}\, x_1 - x_1\, x_2,\; \alpha^{-1}\, \mathrm{adj}\, x_2 - x_0\, x_1,\; \alpha\, \mathrm{adj}\, x_1 - x_2\, x_0),$

(16) $\quad \sigma(x, y) = \tau(x_0\, y_0 + x_1\, y_2 + x_2\, y_1).$

Define a rational map $j: V \to V$ by

$$jx = N(x)^{-1} \cdot n(x) \quad (\text{if } N(x) \neq 0).$$

We claim that (V, j, e) is a J-structure. In order to establish this, it suffices by 5.5 to show that:
(a) N is irreducible,
(b) the formulas (1), (2), (3) and (4) hold.

We know that v is a irreducible polynomial function on A (see 2.5). Since v can be identified with the restriction of N to the subspace $(A, 0, 0)$ of V, it follows that N is irreducible. Also, (2) is trivially verified. So to prove that (V, j, e) is a J-structure it remains to be shown that formulas (1), (3) and (4) are verified. This we shall do now.

5.11. For $a, b, c \in A$ define $v(a, b)$ and $v(a, b, c)$ as in 5.2. It follows from 1.11, 1.18 and 2.5 that

(17) $\quad v(1, a) = \tau(a),$

(18) $\quad v(1, a, b) = \tau a \cdot \tau b - \tau(ab).$

The definition (14) of N then shows that we have

(19) $\quad \begin{aligned} N(x, y) &= (dN)_x(y) = (d^2 N)_y(x) \\ &= v(x_0, y_0) + v(x_1, y_1) + v(x_2, y_2) - \tau(y_0\, x_1\, x_2) \\ &\quad - \tau(x_0\, y_1\, x_2) - \tau(x_0\, x_1\, y_2), \end{aligned}$

whence
$$(dN)_e(x) = v(1, x_0),$$
$$(d^2 N)_e(x) = v(x_0, 1) - \tau(x_1 \, x_2),$$
$$(d^2 N)_e(x, y) = v(x_0, y_0, 1) - \tau(x_1 \, y_2) - \tau(y_1 \, x_2).$$

It follows that
$$N(e, x) N(e, y) - N(e, x, y)$$
$$= v(1, x_0) v(1, y_0) - v(1, x_0, y_0) + \tau(x_1 \, y_2 + x_2 \, y_1).$$

By (16), (17) and (18) the right-hand side equals $\sigma(x, y)$. This shows that (1) holds.

(4) follows from (19) and the definitions (15) and (16) of n and σ, bearing in mind that
$$v(a, b) = \tau(\mathrm{adj}\, a \cdot b) \quad (a, b \in A),$$
which is obvious in the matrix algebra $\mathbb{M}_3$.

5.12. It finally remains to prove (3). This can be done by a straightforward calculation, to be found in [23, p. 508]. A less computational proof is as follows.

We assume $A = \mathbb{M}_3$. Let a_0, a_1, a_2 be invertible elements of A with $v(a_1) = 1$.
Define linear transformations g, g' of V by
$$g(x_0, x_1, x_2) = (a_0 \, x_0 \, a_1^{-1}, a_1 \, x_1 \, a_2^{-1}, a_2 \, x_2 \, a_0^{-1}),$$
$$g'(x_0, x_1, x_2) = (a_1 \, x_0 \, a_0^{-1}, a_0 \, x_1 \, a_2^{-1}, a_2 \, x_2 \, a_1^{-1}).$$

g and g' are nonsingular. It follows from the definition of g and g' that

(20) $\quad \begin{cases} N(g\,x) = N(g'\,x) = N(x) \\ n(g\,x) = g'\,n(x) \end{cases}$

(for the second formula observe that $\mathrm{adj}(a\,x\,b) = b^{-1} \cdot \mathrm{adj}\, x \cdot a^{-1}$, if a and b are invertible elements of A with $v(a) = v(b) = 1$).

We now claim: *there is an open subset $U \ne \emptyset$ of V such that for $x = (x_0, x_1, x_2) \in U$ there exists g with $g\,x = (y_0, y_1, y_2)$, where the y_i are nonsingular diagonal matrices.*

Let U be the subset of the $x = (x_0, x_1, x_2) \in V$ such that $x_0 \, x_1 \, x_2$ is a nonsingular matrix with three distinct eigenvalues. Clearly U is open and nonempty. If $x \in U$, let $a_0 \in A$ be such that $v(a_0) = 1$ and that $a_0 \, x_0 \, x_1 \, x_2 \, a_0^{-1}$ is a diagonal matrix. Taking for a_1 and a_2 suitable scalar multiples of $a_0 \, x_0$ and $a_0 \, x_2^{-1}$, the corresponding g will be as required.

§5. J-structures of Low Degree

Now (3) is easily checked if $x = (x_0, x_1, x_2)$ with diagonal x_i. It then follows form (20) that this formula also holds if $x \in U$. Since U is open and nonempty, it follows by continuity that (3) holds for all $x \in V$.

5.13. We denote the J-structure introduced in 5.10 by $\mathscr{E}_3(A, \alpha)$ and we call it *an exceptional J-structure of the first kind*. If A is defined over $k \subset K$ and if $\alpha \in k^*$ it follows readily from the definitions that $\mathscr{E}_3(A, \alpha)$ is defined over k.

5.14 Lemma. *Let A be defined over $k \subset K$, let $\alpha, \beta \in k^*$. Assume that there exists $u \in A(k)$ such that $\beta = \alpha \, \nu(u)$. Then $\mathscr{E}_3(A, \alpha)$ and $\mathscr{E}_3(A, \beta)$ are k-isomorphic.*

The underlying vectorspaces of $\mathscr{E}_3(A, \alpha)$ and $\mathscr{E}_3(A, \beta)$ are the same, viz. $A \oplus A \oplus A$. It is readily checked that the automorphism $(x_0, x_1, x_2) \mapsto (x_0, x_1 u, u^{-1} x_2)$ yields an isomorphism of $\mathscr{E}_3(A, \beta)$ onto $\mathscr{E}_3(A, \alpha)$.

Put $\mathscr{E}_3 = \mathscr{E}_3(\mathbb{M}_3, 1)$. It follows from 5.14 that $\mathscr{E}_3(A, \alpha)$ is K-isomorphic to $\mathscr{E}_3$ (but this need not to be so over $k \subset K$, if $\mathscr{E}_3(A, \alpha)$ is defined over k).

It is obvious from (16) that the symmetric bilinear form σ is nondegenerate. Hence by 5.8 the inner structure group of $\mathscr{E}_3$ acts irreducibly.

$\mathscr{E}_3$ is a J-structure of degree 3. The dimension of its underlying vector space is 27. Observe that $\mathscr{E}_3$ is not isomorphic to any of the J-structures $\mathscr{M}_r, \mathscr{S}_r, \mathscr{A}_r, \mathscr{O}_{2,r}, \mathscr{O}'_{2,r}, \mathscr{O}''_{2,r}$, discussed in §2. By degrees, such an isomorphism could only be with $\mathscr{M}_3, \mathscr{S}_3$ or $\mathscr{A}_3$, whose underlying vector spaces have dimensions different from 27 (namely 9, 6 and 15).

5.15 Exceptional J-structures of the second kind. Let B be an algebra with an involution of the second kind ρ which is simple over K (as an algebra with involution). This means that there is a simple algebra A over K and an involution ι of A such that $B = A \oplus A$ and that $\rho(a_1, a_2) = (\iota a_2, \iota a_1)$.

Assume that A is of degree 3, so that A is isomorphic to $\mathbb{M}_3$. Put

$$B_0 = \{x \in B \mid \rho x = x\},$$
$$V = B_0 \oplus B, \quad e = (1, 0).$$

B consists of the elements $(a, \iota a)$ $(a \in A)$.

Let again ν and τ denote reduced norm and reduced trace of A. Put $L = K \oplus K$, this is the centre of B. We consider K as the subfield of ρ-invariant elements of L. Define $\mu: B \to L$ by $\mu(a_1, a_2) = (\nu(a_1), \nu(a_2))$. Since $\mu(\rho x) = \mu x$ for $x \in B_0$, we have a polynomial map $\mu_0: B_0 \to K$ with $\mu_0(a, \iota a) = \nu(a)$. Let α be an invertible element of L, let $u \in B_0$ be such that $\mu_0(u) = \alpha \cdot \rho \alpha$. Denote by t the linear map $B \to L$ with $t(a_1, a_2) = (\tau a_1, \tau a_2)$, let t_0 be the linear map $B_0 \to K$ with $t(a, \iota a) = \tau a$.

Let q, q_0 be the quadratic maps $B \to B$ and $B_0 \to B_0$, respectively, with
$$q(a_1, a_2) = (\text{adj } a_1, \text{adj } a_2)$$
$$q_0(a, a) = (\text{adj } a, \iota(\text{adj } a)).$$

We now define the ingredients N, n and σ for a J-structure by the following formulas (in which $v = (x_0, x)$ and $w = (y_0, y)$ are elements of $V = B_0 \oplus B$).

(21) $\qquad N(v) = \mu_0(x_0) + \alpha \mu(x) + \rho \alpha \cdot \mu(\rho a) - t(x_0 \, x \, u(\rho \, x)),$

(22) $\qquad n(v) = (q_0 x_0 - x u(\rho x), \rho \alpha \cdot q(\rho \, x) u^{-1} - x_0 \, x),$

(23) $\qquad \sigma(v, w) = t_0(x_0 y_0) + t_0(x u(\rho y) + y u(\rho x)).$

Define the rational map $j: V \to V$ by
$$j x = N(x)^{-1} \cdot n(x) \qquad (N(x) \neq 0).$$

We claim that (V, j, e) is a J-structure. To show this, let $\alpha = (\alpha_1, \alpha_2)$, $u = (u_1, \iota u_1)$. Define a linear isomorphism $\phi: A \oplus A \oplus A \to V$ by
$$\phi(x_0, x_1, x_2) = ((x_0, \iota x_0), (x_1, \iota(u_1^{-1} x_2))).$$

If we express $N(\phi x)$, $n(\phi x)$, $\sigma(\phi x, \phi y)$ in terms of x_0, x_1, x_2, we get the formulas (14), (15), (16), with α replaced by α_1. This establishes our claim.

5.16. We denote the J-structure defined in 5.6 by $\mathscr{E}_3(B, \rho, u, \alpha)$ and we call it *an exceptional J-structure of the second kind*. From what we established in 5.15 it is clear that this J-structure is isomorphic to $\mathscr{E}_3$. However, if we take into account fields of definition the construction of 5.15 becomes more interesting.

Let l and k be subfields of K, such that l is separable quadratic extension of k. Let C be a central simple algebra of degree 3 over l, with an involution τ of the second kind, whose field of invariants in the centre l is k. Put $B = C \otimes_k K$, $\sigma = \tau \otimes \text{id}$. Let $\alpha \in l^*$, $u \in C(k)$ such that $\tau u = u$, $\mu(u) = \alpha \cdot \text{id}$ (where μ denotes the restriction of the reduced norm of C).
Then the J-structure $\mathscr{E}_3(B, \rho, u, \alpha)$ is defined over k, as follows readily form (21) and (22). Such a J-structure is not necessarily k-isomorphic to an exceptional J-structure of the first kind.

Notes

The treatment of J-structures of degree 3 given here is similar to that given by McCrimmon of exceptional Jordan algebras in [23] and [24]. 5.5 is a variant of [23, Theorem 1, p. 499]. The usefulness of results like 5.7 in characteristic 2 will appear in §7.

§5. J-structures of Low Degree

Tits' construction of exceptional Jordan algebras is discussed in [14, p. 412-414], if $\text{char}(K) \neq 2$ and in [23] and [24]. We shall return to the construction in §15.

It will be shown in the next sections that there is a close connection between J-structures and Jordan algebras. The exceptional J-structures are then related to the so-called exceptional Jordan algebras. The usual way of introducing these algebras is by using hermitian 3×3 matrices over an octave algebra (or Cayley algebra). This suggest another way of defining $\mathscr{E}_3$, taking V to be the space of these 3×3 matrices. We do not discuss the details of this alternative approach to $\mathscr{E}_3$ here. The interested reader will have no difficulty in extracting the details from the literature on Jordan algebras.

In particular, the formulas which are relevant for the definition can be found in [23, p. 501-502]. The proof that the conditions of 5.5 are verified can also be based on the results established there. The connection between the two approaches to $\mathscr{E}_3$ is discussed in [14, p. 414-415]. This alternative approach to $\mathscr{E}_3$ will not be used in the sequel.

§6. Relation with Jordan Algebras (Characteristic $\neq 2$)

In this section we shall show that if $\operatorname{char}(K)\neq 2$, a J-structure is essentially the same thing as a Jordan algebra. We denote by A a finite dimensional commutative algebra with identity element e. A is not assumed to be associative. The linear transformation $y \mapsto xy$ of A is denoted by $L(x)$.

6.1 Lemma. (i) *There exists a unique birational map* $i: A \to A$ *such that* $ix \cdot x = e$ *if i is regular in* $x \in A$;
(ii) *i is homogeneous of degree -1 and $i^2 = \operatorname{id}$, i is regular in e and $ie = e$*;
(iii) *If A is defined over $k \subset K$ then i is defined over k.*

The proof is similar to that of 2.1. $L(x)$ is nonsingular for x in a nonempty open subset U of A, containing e. Defining $ix = L(x)^{-1} \cdot e$ $(x \in U)$ it is clear that i satisfies (i).

The uniqueness follows from the fact that the requirement of (i) defines i on the open subset U of A. (ii) and (iii) are trivial (to prove $i^2 = \operatorname{id}$ one needs the commutativity of A).

We call the birational map i the *inversion* of A. We say that $x \in A$ is *invertible* if i is regular in x and we then write $x^{-1} = ix$. We call x^{-1} the *inverse* of x.

6.2. We define the powers of an element $x \in A$ as in 3.10. Let X be an indeterminate over K. Then there is a formal power series

(1) $$i(e - xX) = \sum_{m=0}^{\infty} x^m X^m,$$

where $x \mapsto x^m$ is a rational map $A \to A$. From

$$(e - xX) i(e - xX) = e$$

we obtain that $x^0 = e$, $x^1 = x$ and that

(2) $$x^{m+1} = x \cdot x^m \quad (m \geq 0),$$

from which one sees that $x \mapsto x^m$ is a polynomial map of degree m. We call x^m the m-th power of x. (2) shows that our definition of powers is the usual one.

§6. Relation with Jordan Algebras (Characteristic $\neq 2$)

6.3 Jordan algebras. We recall that A is said to be a *Jordan algebra* if we have

(3) $\qquad x(x^2 y) = x^2(x y) \qquad (x, y \in A).$

From (3) one finds, replacing x by $x+z$ and comparing linear terms in z in both sides of the resulting equality,

$$2x((x z) y) + z(x^2 y) = 2(x y)(x z) + x^2(y z),$$

whence

$$L(x^2 y) = -2L(x) L(y) L(x) + 2L(x y) L(x) + L(x^2) L(y) \qquad (x, y \in A).$$

Putting $y = x^{m-1}$ ($m \geq 2$) and using that $x^2 \cdot x^{m-1} = x^{m+1}$ (which follows from (2) and (3)) this gives that

$$L(x^{m+1}) = -2L(x) L(x^{m-1}) L(x) + 2L(x^m) L(x) + L(x^2) L(x^{m-1}).$$

By induction one then proves that

(4) $\qquad L(x) L(x^m) = L(x^m) L(x) \qquad (m \geq 1).$

We can now prove another characterization of Jordan algebras.

6.4 Proposition. *A is a Jordan algebra if and only if we have for all invertible $x \in A$ and all $y \in A$ that*

(5) $\qquad x^{-1}(x y) = x(x^{-1} y).$

Suppose that A is a Jordan algebra. Let X be an indeterminate over K. Using (1) and (4) we obtain that the formal power series for $(e - xX) \cdot (i(e - xX) y)$ and $i(e - xX)((e - xX) y)$ are equal.

It follows that we have

$$(e - t x)(i(e - t x) y) = i(e - t x)((e - t x) y),$$

if $x, y \in A$ and $t \in K$ such that $e - t x$ is invertible. This implies (5) for x in a suitable nonempty open subset of A. By continuity it then follows that (5) holds for all invertible x.

Conversely, assume that (5) holds. Replacing x by $e - xX$ in (5) and using (1) we find the defining Jordan algebra identity $x(x^2 y) = x^2(x y)$ by equating coefficients of X^3 in both sides of the resulting equality.

The next theorem is the main result of this section.

6.5 Theorem. *Assume that* $\text{char}(K) \neq 2$.
(i) *Let A be a Jordan algebra with inversion i and identity e. Then (A, i, e) is a J-structure;*
(ii) *Let (V, j, e) be a J-structure. There exists a unique Jordan algebra structure on V with identity element e, whose inversion is j.*

Proof of (i). We have to establish that the axioms (J1), (J2) and (J3) of 1.3 hold for (A, i, e). That (J1) holds follows from 6.1 (which holds for any commutative algebra with identity).

We next prove (J2). It follows from (5), by induction on m, that

(6) $$x^{-1} \cdot x^m = x^{m-1},$$

if $x \in A$ is invertible. Let X be an indeterminate over K. It follows from (6) that we have the following identity of formal power series

$$(X - x^{-1}) \sum_{m=1}^{\infty} x^m X^m = -X.$$

By (1) this implies that

$$(X - x^{-1})(e - i(e - xX)) = -X,$$

from which we conclude that there is a nonempty subset U of A such that

$$(e - ix)(e - i(e - x)) = -e,$$

if $x \in U$. This implies (J2): the last formula is easily seen to be equivalent to §1, (3).

It remains to prove (J3). If $x \in A$ is invertible, let

$$P(x)^{-1} = -(di)_x.$$

From $x \cdot x^{-1} = e$ we obtain by differentiation that

$$x(di)_x(y) + y x^{-1} = 0,$$

whence

(7) $$L(x^{-1}) P(x) = L(x).$$

From $x^{-1} \cdot x^2 = x$ (which is a particular case of (6)) we find by differentiation that

$$x^{-1} \cdot 2xy + (di)_x(y) x^2 = y.$$

Using (5) and (7) this implies that

(8) $$P(x) y = 2x(x y) - x^2 y.$$

It follows that the rational map $x \mapsto P(x)$ of A into the space of endomorphisms $\text{End}(A)$ is a quadratic polynomial map. Put

$$P(x, y) = (dP)_x(y) \quad (x, y \in A).$$

$P(x, y)$ is a linear transformation of A, which by (8) is given by

(9) $$P(x, y) z = 2x(y z) + 2y(x z) - 2z(x y) \quad (x, y, z \in A).$$

§6. Relation with Jordan Algebras (Characteristic $\neq 2$)

(9) and (5) imply

(10) $$P(x, y) y^{-1} = 2x,$$

if y is invertible.

Differentiation of (7) gives that
$$-(P(x)^{-1} y)(P(x) z) + x^{-1}(P(x, y) z) = y z.$$
Taking $z = y^{-1}$ and using (10) we get
$$P(x)^{-1}(y) \cdot (P(x) y^{-1}) = e,$$
if x and y are invertible. But this means that all $P(x)$ (x invertible) are in the structure group of i. To establish (J 3) it now suffices to prove that the set of elements $P(x) e$ (x invertible in A) contains a nonempty open subset of A. But by (8) we have $P(x) e = x^2$. 2.9 now gives what we want (observe that the hypothesis $\operatorname{char}(K) \neq 2$ is needed here). This finishes the proof of (i).

Proof of (ii). We use the notations of §3 for the J-structure (V, j, e). If we have an algebra structure on V such that $jx = x^{-1}$ for all invertible elements (for the algebra structure) then comparing (1) with the definition of powers in a J-structure, given in §3, (17), we see that the square of an element of the algebra must be the same as the square in the J-structure. P now being as in §3, we have $x^2 = P(x) e$. Replacing x by $x + y$, it follows that the product of the algebra structure must be given by

(11) $$2xy = P(x, y) e.$$

This already proves the uniqueness statement of (ii). To prove that (11) defines a Jordan algebra structure, it suffices by 6.4 to prove (5). Let j be regular in x. From §3, (8) and (10) it follows that
$$P(x, j x) e = P(x, e) j x = 2 e.$$
(11) then shows that
$$x \cdot j x = e.$$

Consequently j is inversion of the algebra structure defined by (11). We have remarked in §3 (immediately after §3, (13)) that $P(x, e)$ and $P(j x, e)$ commute, hence
$$P(j x, e) P(x, e) y = P(x, e) P(j x, e) y.$$
By §3, (10) this is the same as (5). Hence (11) defines a Jordan algebra structure.

6.6 Corollary. *Let A be a Jordan algebra with inversion i. Then we have, if $x \in A$ is invertible,*
$$-(di)_x^{-1} y = 2 x(x y) - x^2 y.$$

This is formula (8), which was established in the course of the proof of (i).

6.7. Let A be a Jordan algebra (in characteristic $\neq 2$) with identity element e and inversion i. We denote the J-structure (A, i, e) by $\mathscr{J}(A)$ and we call it *the J-structure defined by the Jordan algebra* A. If A is a commutative associative algebra then this J-structure is the same as that of 2.1.

If A is an arbitrary associative algebra and if $\operatorname{char}(K) \neq 2$ define a commutative multiplication $x \circ y$ on A by

$$2 x \circ y = (xy + yx).$$

Let A^+ denote the resulting algebra. It is well-known (and easy to prove) that A^+ is a Jordan algebra. It then follows that the J-structure $\mathscr{J}(A)$ of 2.1 coincides with the J-structure $\mathscr{J}(A^+)$ of 6.5(i).

It should be remarked that the correspondence between J-structures and Jordan algebras, given in 6.5, possesses the obvious functorial properties.

6.8 Invariant bilinear forms of $\mathscr{J}(A)$. Let A be a Jordan algebra. With the notations of 6.7 let B be an invariant bilinear form in $V \times V$, with respect to $\mathscr{J}(A)$ (see 1.22).

From §1, (19) and 1.16(v) we find that

(12) $$B(P(x) y, z) = B(y, P(x) z),$$

whence
$$B(P(x, e) y, z) = B(y, P(x, e) z).$$

By (11) this means that

(13) $$B(x y, z) = B(y, x z).$$

In particular, if σ is the standard symmetric form of $\mathscr{J}(A)$ and τ the trace of $\mathscr{J}(A)$ (see 1.11) then
$$\sigma(x, y) = \tau(x y).$$

Conversely, if (13) holds then (8) implies that (12) holds and it follows that B satisfies §1, (19) for all g in the inner structure group.

6.9 The structure algebra of $\mathscr{J}(A)$. In the situation of 6.7, let $\mathfrak{g}$ and $\mathfrak{h}$ be the structure algebra and the derivation algebra of the J-structure $\mathscr{J}(A)$ (these Lie algebras were defined in §4). We shall give a more explicit description of $\mathfrak{g}$ and $\mathfrak{h}$.

6.10 Proposition. (i) $\mathfrak{h}$ *consists of the linear transformations X of A such that*

(14) $$X(a b) = a(X b) + (X a) b \quad (a, b \in A);$$

(ii) *All $L(a)$ ($a \in A$) lie in $\mathfrak{g}$ and $\mathfrak{g}$ is the vector space direct sum of $\mathfrak{h}$ and $L(A)$.*

§6. Relation with Jordan Algebras (Characteristic $\neq 2$)

By 4.7(i) it follows that $X \in \mathfrak{h}$ if and only if $Xe = 0$ and

$$P(a) X(a^{-1}) = -Xa \quad (a \text{ invertible}),$$

where P is the quadratic map of $\mathcal{J}(A)$.

Let T be an indeterminate over K. Replacing a by $e - aT$ in the last formula it follows from the resulting formula that $X \in \mathfrak{h}$ if and only if $Xe = 0$ and

(15) $\quad X a^n = P(e, a) X a^{n-1} - P(a) X a^{n-2} \quad (n \geq 2).$

In particular,

$$X a^2 = P(e, a) X a = 2 a \cdot X a,$$

which implies (14). Conversely, (14) implies $Xe = 0$ and (15) (use (8)). This establishes (i).

From (4) and (8) it follows that $P(x)$ and $L(x)$ commute. Then (7) implies that $P(x) L(x^{-1}) = L(x)$. Hence

$$P(x)(a x^{-1}) = a x \quad (a \in A, x \text{ invertible}).$$

By §4, (2) this shows that $L(a) \in \mathfrak{g}$.

Let $X \in \mathfrak{g}$. Then it follows, using (i), that $X - L(Xe) \in h$. Hence $\mathfrak{g} = \mathfrak{h} + L(A)$. That this is a direct sum is obvious.

6.11. Let $L(a, b)$ be as in §4. By §4, (4) and (9) we have

$$L(a, b) x = 2 a(b x) + 2(a b) x - 2 b(a x).$$

From this formula one infers, using 6.10(i) that the Lie algebra $\mathfrak{h}_1$ of inner derivations of $\mathcal{J}(A)$ now consists of all sums of linear transformations of A of the form

$$x \mapsto a(b x) - b(a x),$$

which is the usual definition of inner derivations of a Jordan algebra, see [14, p. 35].

Notes

The symmetrization of an associative multiplication, briefly mentioned in 6.7, was the procedure which led P. Jordan to the identity (3). A result of the same type as 6.4 was first proved by Koecher [16, Satz 1, p. 70]. The deduction of (J3), given in the proof of 6.5, is taken from [8, p. 66–69]. 6.10 is a familiar result, see [8, IX, 5.1, p. 289].

§7. Relation with Quadratic Jordan Algebras

In the main result of §6 (Theorem 6.5) we had to assume $\operatorname{char}(K) \neq 2$. If one wishes to include the characteristic 2 case, Jordan algebras do not suffice. One then needs the quadratic Jordan algebras. We first recall the definition, in a form adapted to our point of view.

7.1 Quadratic Jordan algebras. A *quadratic Jordan algebra* is a triple $\mathcal{Q} = (V, P, e)$, where V is a finite dimensional vector space over K, P a quadratic map of V into the space $\operatorname{End}(V)$ of its endomorphisms and e a nonzero element of V, such that the axioms to be stated below hold. Putting $P(x, y) = P(x+y) - P(x) - P(y)$ we obtain a symmetric bilinear map $P(\ ,\ )$ of $V \times V$ into $\operatorname{End}(V)$.

The axioms for a quadratic Jordan algebra are as follows:
(QJ1) $P(e) = e$, $P(x, e) y = P(x, y) e$;
(QJ2) $P(P(x) y) = P(x) P(y) P(x)$;
(QJ3) $P(x) P(y, z) x = P(P(x) y, x) z$.

$\mathcal{Q}$ is said to be *defined over* $k \subset K$ if there exists a k-structure on the vector space V such that P is defined over k and that $e \in V(k)$.

Let $\mathcal{Q} = (V, P, e)$ be a quadratic Jordan algebra.

7.2 Lemma. (i) *There exists a unique birational map* $i \colon V \to V$ *such that* $P(x) i x = x$ *if* x *is regular in* $x \in V$;
(ii) i *is homogeneous of degree* -1 *and* $i^2 = \operatorname{id}$, i *is regular in* e *and* $i e = e$;
(iii) *If* $\mathcal{Q}$ *is defined over* $k \subset K$ *then* i *is defined over* k.

This is again a result like 2.1 and 6.1. Let U be the set of $x \in V$ such that $P(x)$ is invertible. U is open and nonempty. For $x \in U$ define $i x = P(x)^{-1} \cdot x$. Then i is as required. We shall only prove that $i^2 = \operatorname{id}$, which is the least trivial assertion of 7.2. By (QJ2) we have $P(x) = P(x) P(i x) P(x)$, hence $P(i x) = P(x)^{-1}$ ($x \in U$). Then $i^2 x = P(i x)^{-1} \cdot i x = P(x) i x = x$ ($x \in U$).

We call the birational map i the *inversion* of $\mathcal{Q}$. $x \in V$ is called *invertible* if i is regular in x and $i x$ is called its *inverse*.

7.3 Lemma. *Let x be invertible. Then*

(1) $$P(i x) = P(x)^{-1},$$

§7. Relation with Quadratic Jordan Algebras

(2) $\qquad P(x,y)\,i\,x = 2\,y \qquad (y \in V),$

(3) $\qquad i(P(x)\,y) = P(i\,x)\,i\,y \qquad \text{if } y \text{ is invertible.}$

That (1) holds was already remarked above. To prove (2), put $y = i\,x$ in (QJ 3). We then obtain

$$P(x)\,P(i\,x, z)\,x = P(P(x)\,i\,x, x)\,z = P(x, x)\,z = 2\,P(x)\,z,$$

which implies (2).

It follows from (QJ 2) that

$$P(P(x)\,y)(P(i\,x)\,i\,y) = P(x)\,P(y)\,P(x)\,P(i\,x)\,i\,y = P(x)\,y,$$

whence (3) by 7.2(i).

7.4. We next define the powers of an element $x \in V$. Let X be an indeterminate over K. We then have a formal power series

(4) $\qquad i(e - x\,X) = \sum_{m=0}^{\infty} x^m\,X^m,$

where $x \mapsto x^m$ is a rational map $V \to V$.

7.5 Lemma. (i) $x \mapsto x^m$ *is a homogeneous polynomial map of degree* m, *we have* $x^0 = e$, $x^1 = x$;
(ii) *The following relations hold for* $x \in V$, $i, j, m \geq 0$

(5) $\qquad P(x)\,x^m = x^{m+2},$

(6) $\qquad P(x^i, x^j)\,x^m = 2\,x^{i+j+m}.$

The proof of this lemma is similar to that of 3.11 and is therefore omitted. In the proof the formulas (QJ 1), (QJ 2) and (2) are used.

x^m is defined for all $x \in V$ (by 7.5(i)), it is called the m-th power of x. $x^2 = P(x)\,e$ is called the square of x.

It might be expected that (V, i, e) is a J-structure if $\mathscr{Q} = (V, P, e)$ is a quadratic Jordan algebra. But this is not so.

7.6 Example. Let $\operatorname{char}(K) = 2$. Let V be a finite dimensional vector space and Q a quadratic form on V. Let $e \in V$ be such that $Q(e) = 1$. We use the notations of 2.15. Put

(7) $\qquad P(x)\,y = Q(x, \bar{y})\,x - Q(x)\,\bar{y} \qquad (x, y \in V).$

It can then be verified that $\mathscr{Q} = (V, P, e)$ is a quadratic Jordan algebra. Putting (as in §2)

$$j\,x = Q(x)^{-1} \cdot \bar{x}$$

it follows from 7.2(i) and (7) that j is the inversion in $\mathscr{Q}$. As in §2 one sees that the structure group of j is the group G of all $g \in GL(V)$ which leave Q invariant up to a nonzero scalar factor.

Suppose now that (with the notation of 2.17) $Q_e = 0$, so that 2.17 does not apply to (V, j, e). In that case we have that $Q_x = 0$ for all x in the orbit Ge of e under the structure group G. It follows that if $Q_e = 0$ but $Q_x \neq 0$ for some $x \in V$, (V, j, e) is *not* a J-structure because axiom (J3) of 1.3 does not hold.

To obtain a connection between quadratic Jordan algebras and J-structures in characteristic 2, we shall have to make a restrictive assumption.

7.7 Definition. *Let* char$(K) = 2$. *Let $\mathscr{A}$ be either a J-structure or a quadratic Jordan algebra, with underlying vector space V. $\mathscr{A}$ is said to be separable if the squaring map $x \mapsto x^2$ of V is dominant (i.e. if its image is dense in V).*

7.8 Examples. We assume char$(K) = 2$.
(a) Let A be a finite dimensional commutative and associative algebra over K. The squaring map $s: x \mapsto x^2$ of A is now a semilinear map of K (with respect to the squaring isomorphism of K), hence $s(A)$ is a subspace of A. s is dominant if and only if s is surjective, which is the case if and only if Ker $s = 0$. But Ker $s = 0$ means that A is reduced, i.e. has no nontrivial nilpotent elements. It follows that the J-structure $\mathscr{J}(A)$ of 2.1 is separable if and only if A is reduced.

If moreover A is defined over $k \subset K$, then A is separable if and only if $A(k) \otimes_k K$ is reduced. As is well-known, this means that $A(k)$ is a direct sum of a finite number of separable field extensions of k.
(b) The example (a) also shows that nonseparable J-structures exist, for example the J-structure $\mathscr{J}(K[\delta])$, where $K[\delta]$ is the K-algebra of dual numbers.
(c) In the situation of (a), put $P(x) y = x^2 y$ $(x, y \in A)$. e denoting the identity element of A, it is trivially verified that (A, P, e) is a nonseparable quadratic Jordan algebra.

We next discuss separability of the examples of §2 and §5.

7.9 Proposition. *Let* char$(K) = 2$. *The J-structures $\mathscr{M}_r$, $\mathscr{S}_r$, $\mathscr{A}_r$, $\mathscr{O}'_{2,r}$, $\mathscr{O}''_{2,r}$ and $\mathscr{E}_3$ are separable.*

We discuss the individual cases.

$\mathscr{M}_r$. The underlying vector space of the J-structure is the space $\mathbb{M}_r$ of $r \times r$ matrices. The square of a matrix $X \in \mathbb{M}_r$ in the sense of J-structures is the usual square. Let U be the set of matrices whose eigenvalues are all distinct. U is a nonempty open subset of $\mathbb{M}_r$.

If $X \in U$, the subalgebra $K[X]$ of $\mathbb{M}_r$ generated by X is a direct sum of r copies of K. It follows that there exists $Y \in K[X]$ such that $Y^2 = X$.

§7. Relation with Quadratic Jordan Algebras

Hence all elements of U are squares, which shows that $\mathcal{M}_r$ is separable.

$\mathcal{S}_r$. The underlying vector space is the space $\mathbb{S}_r \subset \mathbb{M}_r$ of symmetric matrices. Since $\mathcal{S}_r$ is a substructure of $\mathcal{M}_r$ (see 2.10) the square of $X \in \mathbb{S}_r$ is again the usual square.

With the previous notations, $U \cap \mathbb{S}_r$ is open in $\mathbb{S}_r$ (and nonempty). If $X \in U \cap \mathbb{S}_r$ and if $Y \in K[X]$ is such that $Y^2 = X$, then clearly $Y \in \mathbb{S}_r$. It follows that all elements of $U \cap \mathbb{S}_r$ are squares. Hence $\mathcal{S}_r$ is separable.

$\mathcal{A}_r$. The underlying vector space is the space $\mathbb{A}_{2r} \subset \mathbb{M}_{2r}$ of alternating $2r \times 2r$ matrices. Let $S \in \mathbb{A}_{2r}$ be as in 2.13. From the definition of $\mathcal{A}_r$ in §2 one finds, using §3, (17), that the square in $\mathcal{A}_r$ of $X \in \mathbb{A}_{2r}$ is given by $s(X) = XSX$.

Let $U \subset \mathbb{A}_{2r}$ be the subset consisting of the $X \in \mathbb{A}_{2r}$ such that the polynomial in the indeterminate T given by $\text{Pf}(TS + X)$ (where Pf denotes the Pfaffian, see 2.12) has r distinct roots. U is clearly open. U is nonempty, since it contains the matrices $X = (x_{ij})$ with $x_{ij} = 0$ if $|i-j| \neq 1$, $x_{i,i+1} = x_{i+1,i} = x_i$, where the $x_i \in K$ are all distinct and not equal to 1.

Let $X \in U$. It then follows from 2.12(i) that the characteristic polynomial of SX is the square of a polynomial of degree r, with r distinct roots. We claim that SX is a semisimple (= diagonalizable) matrix. In order to show this, let $V = K^{2r}$, let $(e_i)_{1 \leq i \leq 2r}$ be the canonical basis of V. Define the symmetric bilinear form $(\ ,\ )$ on $V \times V$ by

Put
$$\left(\sum_{i=1}^{2r} x_i e_i, \sum_{i=1}^{2r} y_i e_i \right) = \sum_{i=1}^{2r} x_i y_i.$$

$$F(x, y) = (x, Sy), \quad G(x, y) = (x, Xy).$$

F and G are alternating bilinear forms on $V \times V$. F is nondegenerate (since S is nonsingular) and we have

$$G(x, y) = F(x, SXy).$$

Suppose that SX is not semisimple. Then there exist linearly independent $x, y \in V$ and $a \in K$, such that

Then
$$SXx = ax, \quad SXy = ay + x.$$

$$0 = G(y, y) = F(y, SXy) = aF(y, y) + F(y, x) = F(x, y).$$

Also, if $z \in V$, $b \neq a$ and $(SX - b)^n z = 0$, we have

$$0 = F((SX - b)^n z, x) = F(x, (SX - b)^n x) = (a - b)^n F(z, x),$$

whence $F(z, x) = 0$. Similarly $F(z, y) = 0$.

But it now follows from what we remarked before about the characteristic polynomial of SX, that we have $F(x, z) = 0$ for all $z \in V$. This contradicts the nondegeneracy of F. Hence SX is semisimple.

It follows that there is a splitting $V = \bigoplus_{i=1}^{r} V_i$ into a sum of 2-dimensional subspaces V_i, such that all V_i are stable for SX and that the restriction of SX to V_i is a scalar multiplication, say by a_i. Define a linear transformation Y of V by requiring that all V_i are stable for SY and that SY restricted to V_i is scalar multiplication by $a_i^{\frac{1}{2}}$. Then clearly $(SY)^2 = SX$, hence $X = YSY$. Moreover $Y \in \mathbf{A}_{2r}$.

It follows that all elements of U are squares, which establishes the separability of A_r.

$\mathcal{O}'_{2,r}$ and $\mathcal{O}''_{2,r}$. Let $(V, j, e) = \mathcal{O}'_{2,r}$ (resp. $\mathcal{O}''_{2,r}$). From its definition (see 2.20) it follows that there exists a quadratic form Q on V, with associated symmetric bilinear form $Q(\ ,\ )$, such that the squaring map s is given by

$$sx = Q(x, e)x - Q(x)e.$$

It is now easy to check that every $x \in V$ such that $Q(x, e) \neq 0$ is the square of an element of the form $x + ae$ ($a \in K$). This proves separability.

$\mathcal{E}_3$. In this case the separability is a consequence of 5.7. This completes the proof of 7.9.

We now turn to the connection between J-structures and quadratic Jordan algebras.

7.10 Theorem. (i) Let $\mathscr{S} = (V, j, e)$ be a J-structure. Let P be the quadratic map of $\mathscr{S}$. Then (V, P, e) is a quadratic Jordan algebra;
(ii) Let $\mathscr{Q} = (V, P, e)$ be a quadratic Jordan algebra. If $\operatorname{char}(K) = 2$, assume $\mathscr{Q}$ to be separable. Let i be the inversion of $\mathscr{Q}$. Then (V, i, e) is a J-structure.

Proof of (i). By 3.4, P is a quadratic map $V \to \operatorname{End}(V)$. (QJ1) follows from 1.16(iv) and §3, (10). (QJ2) is §3, (6). To prove (QJ3), replace x by jx in §3, (9) and apply $P(x)$ to both sides. Using (QJ2) the formula of (QJ3) then follows (observe that no separability is needed for this proof).
Proof of (ii). We have to check the axioms (J1), (J2) and (J3) of 1.3. (J1) follows from 7.2. To prove (J3) we observe that by (3) all $P(x)$ with $x \in V$ invertible lie in the structure group of i. It then suffices to show that the squares $x^2 = P(x)e$ of invertible elements of V form a dense subset of V.

If $\operatorname{char}(K) = 2$, this follows from separability. If $\operatorname{char}(K) \neq 2$ we use that the differential of the squaring map $s: x \mapsto x^2$ at e is given by $(ds)_e(x) = P(x, e)e = P(e, e)x = 2x$ (by (QJ1)). Since $(ds)_e$ is surjective, s has a dense image by [1, 17.3, p. 75].

It remains to prove (J2). Let X be an indeterminate over K. There exists a nonempty open subset U of V such that for $x \in U$ we have a formal power series

(8) $$i(eX - ix) = -\sum_{m=0}^{\infty} \phi_m(x) X^m,$$

§7. Relation with Quadratic Jordan Algebras

where ϕ_m is a rational map $V \to V$. We shall prove presently that $\phi_m(x) = x^{m+1}$. Assuming this, it follows from (4) and (8) that

$$i(e-xX)+i(e-i(xX))=e,$$

which implies (J 2).

To prove $\phi_m(x) = x^{m+1}$, we use 7.2(i) to obtain

$$P(eX-ix)i(eX-ix)=eX-ix,$$

which by inserting (8) gives

$$P(ix)\phi_0(x)=ix,$$
$$P(ix)\phi_1(x)-P(ix,e)\phi_0(x)=-e,$$
$$P(ix)\phi_m(x)-P(ix,e)\phi_{m-1}(x)+\phi_{m-2}(x)=0 \quad (m \geq 2).$$

Using 7.2 and 7.3 we obtain from the first relation that $\phi_0(x)=x$. The second relation then gives

$$P(x)^{-1}\phi_1(x)=P(ix,e)x-e.$$

By (QJ 1) and (2), the right-hand side equals e. Hence $\phi_1(x)=x^2$.

From the third relation one obtains, using (1), that

$$\phi_m(x)=P(x)P(ix,e)\phi_{m-1}(x)-P(x)\phi_{m-2}(x).$$

Assume $m \geq 2$, $\phi_{m-1}(x)=x^m$, $\phi_{m-2}(x)=x^{m-1}$. It then follows from (QJ 2), (5) and (6) that $\phi_m(x)=x^{m+1}$, which establishes our claim, by induction on m. This finishes the proof of 7.10.

7.11 Corollary. *Let* char$(K) \neq 2$.
(i) *Let A be a finite dimensional Jordan algebra with identity element e. Put $P(x)y = 2x(xy)-x^2 y$ $(x, y \in A)$. Then (A, P, e) is a quadratic Jordan algebra;*
(ii) *Let (V, P, e) be a quadratic Jordan algebra. Define a product on V by $xy = \frac{1}{2}P(x, y)e$. Then V, equipped with this product, is a Jordan algebra with identity element e.*

(i) follows from 6.5(i) and 7.10(i), taking into account 6.6.
(ii) follows from 7.10(ii) and 6.5(ii), using §6, (11).

7.12 Remark. It is not known to the author whether the separability of $\mathcal{Q}$, assumed in 7.10(ii) in the characteristic 2 case, is really necessary for the proof that (V, i, e) is a J-structure.

We terminate this section with some special results. Let $\mathscr{S}=(V, j, e)$ be a J-structure. We denote by G the structure group of $\mathscr{S}$, by $g \mapsto g'$ its standard automorphism and by P its quadratic map. Let H be the automorphism group of S.

7.13 Lemma. *If* char$(K)=2$, *assume $\mathscr{S}$ to be separable. Then there exists x_0 in the orbit Ge such that:*

(i) *there are only finitely many* $x \in V$ *with* $x^2 = x_0^2$,
(ii) *there are only finitely many* $x \in V$ *with* $P(x) = P(x_0)$.

Under the assumption of the lemma, the squaring map $s: x \mapsto x^2$ in V has a dense image (see the proof of part (ii) of 7.10). Standard results from algebraic geometry (see [1, p. 38-39]) now show that there is a nonempty open subset U of V such that for all $x \in U$ the fiber $s^{-1}(sx)$ is finite. Because Ge is open in V by (J 3), $Ge \cap U$ is nonempty, whence (i). Since $P(x) = P(x_0)$ implies $x^2 = x_0^2$, (ii) follows immediately from (i).

Let H_1 be the subgroup of G consisting of the $g \in G$ with $g = g'$. Since $g \mapsto g'$ is an automorphism of the algebraic group G it follows that H_1 is a *closed* subgroup of G. Clearly $H \subset H_1$.

7.14 Proposition. *If* $\mathrm{char}(K) = 2$ *let* $\mathscr{S}$ *be separable. Then* H *is of finite index in* H_1.

Put $f(g) = g(g')^{-1}$. f is a morphism $G \to G$. The fibres of f are the cosets of H_1, hence their dimension equals dim H_1. On the other hand we have by 1.16(i) that $f(g) = P(ge)$.

It then follows from 4.6 and 7.13 that there exists $g_0 \in G$ such that the fiber $f^{-1}(fg_0)$ consists of finitely many cosets of g_0 modulo H, so that the dimension of this fiber equals dim H. It follows that dim H = dim H_1, which proves 7.14.

7.15 A restricted Lie algebra in characteristic 2. Let $\mathrm{char}(K) = 2$. Let $\mathscr{Q} = (V, P, e)$ be a quadratic Jordan algebra. We use the notions of 7.1. Put

$$[x, y] = P(x, y) e.$$

x^2 being as in 7.4, it then follows from (QJ 3) that we have

$$[x^2, y] = [x [x y]],$$

from which one infers Jacobi's identity

$$[x [y z]] + [y [z x]] + [z [x y]] = 0.$$

This means that V, equipped with this product, becomes a restricted Lie algebra, (see [26, p. 7]). We shall make use of this Lie algebra in the proof of 10.20(i).

Notes

Quadratic Jordan algebras were introduced by K. McCrimmon [21]. For a thorough discussion see [15]. The notion of separability in characteristic 2, introduced in 7.7, seems to be needed for our treatment of the connection between J-structures and quadratic Jordan algebras. 7.11 is due to McCrimmon [21, p. 1073].
In 10.34 we shall prove, as a complement to 7.13(ii), that in a separable simple J-structure $P(x) = \mathrm{id}$ implies that $x = \pm e$.

§8. The Minimum Polynomial of an Element

8.1. Let $\mathscr{S}=(V,j,e)$ be a J-structure with norm N. Fix an element $a\in V$. Since the polynomial function $t\mapsto N(te-a)$ $(t\in K)$ is nonzero, there exists a rational map $j_a\colon K\to V$ such that

(1) $$j_a(t)=j(te-a),$$

for t in a suitable open subset of K.

We denote by n_a and N_a a numerator and a denominator of j_a (see 0.5), normalized such that N_a has leading coefficient 1. Let f be the degree of N_a and write

$$N_a(t)=\sum_{i=0}^{f} a_i t^{f-i},$$

with $a_0=1$. We call N_a the *minimum polynomial* of a.

8.2 Lemma. (i) *We have* $n_a(t)=\sum_{m=0}^{f-1}\left(\sum_{i=0}^{m} a_i a^{m-i}\right) t^{f-m-1}$;

(ii) *If m is an integer $\geq f$ then* $\sum_{i=0}^{m} a_i a^{m-i}=0$;

(iii) *f is the smallest integer n such that there exist elements $b_1,\ldots,b_n$ in K with $a^m+\sum_{i=1}^{n} b_i a^{m-i}=0$ for all $m\geq n$.*

Let h be the degree of n_a. Replacing in (1) t by t^{-1} we obtain from the homogeneity of j that the rational map

$$t\mapsto t^{f-h-1}\bigl(t^f N_a(t^{-1})\bigr)^{-1}\cdot\bigl(t^h n_a(t^{-1})\bigr)$$

of K into V is regular for $t=0$, with value e. This implies that $h=f-1$. It then follows that

$$t\mapsto \bigl(t^f N_a(t^{-1})\bigr) j(e-ta)$$

is a polynomial map of degree $f-1$.

Let X be an indeterminate over K. Using the formal power series of §3, (17) for $j(e-aX)$ we find that the power series

$$\left(1+\sum_{i=1}^{f} a_i X^i\right)\sum_{i=0}^{\infty} a^i X^i$$

is a polynomial of degree $f-1$, namely $X^{f-1} n_a(X^{-1})$. (i) and (ii) now readily follow.

Let n and $b_1, \ldots, b_n$ be as in (iii). Then

$$\left(1 + \sum_{i=1}^{n} b_i X^i\right) \sum_{i=0}^{\infty} a^i X^i$$

is a polynomial function, from which it follows that

$$t \mapsto \left(t^n + \sum_{i=1}^{n} b_i a^{n-i}\right) j_a(t)$$

is a polynomial map. This implies $n \geq f$. Since $n \leq f$ is trivially true, the assertion of (iii) follows.

8.3 Proposition. *There exists a nonempty open subset U of V such that for $a \in U$ the minimum polynomial N_a of a is given by*

$$N_a(t) = N(t e - a).$$

This is a consequence of 0.8.

8.4 Example. Let A be a finite dimensional associative algebra with identity e. Let $\mathscr{J}(A)$ be the J-structure defined by A (see 2.1). Let $a \in A$. From 6.2 we know that the usual powers of a are the same as the powers of a in the sense of J-structures. Let m_a be the minimum polynomial of a in the sense of associative algebras, i.e. the polynomial m_a of minimal degree in one indeterminate with leading coefficient 1, such that $m_a(a) = 0$. It then follows from 8.2(iii) that m_a is also the minimum polynomial of a in the J-structure $\mathscr{J}(A)$.

In this example it is obviously true that a relation

$$\sum_{i=0}^{n} b_i a^i = 0$$

with $b_0, \ldots, b_n \in K$ implies

$$\sum_{i=0}^{n} b_i a^{m+1} = 0,$$

for all $m \geq 0$. In the general case we have the following result. We use the previous notations.

8.5 Lemma. *Let $b_0, \ldots, b_n$ be elements of K such that*

(2) $$\sum_{i=0}^{n} b_i a^i = 0.$$

Assume that one of the following conditions is verified:
(a) $\operatorname{char}(K) \neq 2$,
(b) a *is a square in* $\mathscr{S}$.

§8. The Minimum Polynomial of an Element

Then

(3) $$\sum_{i=0}^{n} b_i a^{m+i} = 0$$

for all $m \geq 0$.

In case (b) write $a = b^2$. By §3, (18) and (19) it follows (using induction on i) that $a^i = b^{2i}$. In case (a), (3) follows from (2) by applying $(\frac{1}{2}P(a,e))^m$ to both sides of (2), using §3, (20). In case (b) apply $P(b)^m$ to both sides of (2).

8.6. Let X be an indeterminate over K. We also write N_a for the polynomial $\sum_{i=0}^{f} a_i X^{f-i}$ in $K[X]$. Put

$$B = K[X]/N_a K[X],$$

let b be the canonical image of X in B, then $B = K[b]$. It follows from 8.2(ii) that there is a linear map $\phi: B \to V$ such that $\phi(b^i) = a^i$ ($i \geq 0$). Let $V_a = \phi(B)$ be the subspace of V spanned by the powers of a.

8.7 Proposition. *Assume that one of the conditions* (a), (b) *of* 8.5 *is verified. Then ϕ is injective. If $c \in B$ is invertible, then $\phi(c)$ is invertible and $j(\phi(c)) = \phi(c^{-1})$.*

The injectivity of ϕ follows from 8.5 and 8.2(iii). By 8.2(i), j_a is completely determined by the minimum polynomial N_a. Applying this fact in $\mathscr{S}$ and $\mathscr{J}(B)$, using that a and b have the same minimum polynomial, we find that

(4) $$j(te - a) = \phi((t \cdot 1 - b)^{-1}),$$

for t in a nonempty open subset of K.

If $\operatorname{char}(K) = 2$, the assumption is that a is a square in $\mathscr{S}$. Using §3, (18) and §3, (20), we see that then any element of V_a is a square in $\mathscr{S}$. Let $c \in B$. Applying (4) with $-\phi(c)$ instead of a, we get

$$j(te + \phi(c)) = \phi((t \cdot 1 + c)^{-1}),$$

which implies the last assertion.

8.8. In the situation of 8.7, j induces a birational map of V_a, denoted by $j|V_a$ and it follows that $\mathscr{S}(a) = (V_a, j|V_a, e)$ is a J-structure, isomorphic to $\mathscr{J}(B)$, with $B = K[X]/N_a K[X]$. We call $\mathscr{S}(a)$ the *substructure generated by* a. It is defined only if $\operatorname{char}(K) \neq 2$ or if a is a square.

We say that $a \in V$ is *semisimple* if (i) a is a square in $\mathscr{S}$, (ii) $\mathscr{S}(a)$ is isomorphic to a direct sum of copies of the 1-dimensional J-structure $\mathscr{J}(K)$. The second requirement is equivalent to semisimplicity of the commutative associative algebra B.

The following result now gives a "Jordan decomposition" of elements of a J-structure.

8.9 Proposition. *Let $a \in V$. There exist unique elements $a_s, a_n \in V_a$ such that $a = a_s + a_n$, that a_s is semisimple, a_n is nilpotent and such that a_s and a_n are linear combinations of the powers a^i with $i > 0$. If $\operatorname{char}(K) = p > 0$ then a_s and a_n are also linear combinations of the powers a^{p^i} with $i \geq 0$.*

First assume that $\operatorname{char}(K) \neq 2$ or that a is a square. Then $\mathscr{S}(a)$ is defined. Let B, ϕ and b be as in 8.6. In the associative algebra B one has a decomposition $b = b_s + b_n$ with similar properties. This follows, for example, by applying [1, Prop. 4.2, p. 143] to the linear transformation $x \mapsto bx$ of B. Putting $a_s = \phi(b_s)$, $a_n = \phi(b_n)$ all our requirements are verified.

Next let $\operatorname{char}(K) = 2$ and suppose a is not a square. We apply what we have already proved to $x = a^2$, to get a decomposition $x = x_s + x_n$. x_s being semisimple, there exists a_s in the space spanned by the powers of x_s such that $a_s^2 = x_s$, moreover a_s is unique (since $\operatorname{char}(K) = 2$). Because $x_s \in V_a$ we have that $a_s \in V_a$.

It then follows from §3, (20), that $(a - a_s)^2 = a^2 - a_s^2 = x_n$. Since x_n is nilpotent, we also have that $a_n = a - a_s$ is nilpotent. This proves the existence of the decomposition. The uniqueness follows from the uniqueness of a_s, mentioned before.

To establish the penultimate assertion it suffices to show that we can take a_s to be a linear combination of the x_s^i with $i > 0$, which is easily done. The proof of the last point is also left to the reader.

8.10 Corollary. *Let $\operatorname{char}(K) = 2$. Assume that $\mathscr{S}$ is separable. Then the set of semisimple elements of V contains a nonempty open subset.*

Let $x = x_s + x_n$ be the Jordan decomposition of $x \in V$. It is easily seen that there exists a 2-power a such that $x^a = x_s^a$, for all $x \in V$. 8.10 then follows from the observation that if $\mathscr{S}$ is separable, the morphism $x \mapsto x^a$ is dominant.

Notes

8.3 gives the connection of the norm with "generic minimum polynomials". For a discussion of these in Jordan algebras in characteristic not 2 see [14, Chapter VI]. A discussion of the situation of 8.6 and 8.7, from the point of view of quadratic Jordan algebras, is given in [15, p. 1.61–1.62]. 8.9 is an extension to J-structures of a familiar result in the theory of associative algebras.

§9. Ideals, the Radical

Let $\mathscr{S}=(V,j,e)$ be a J-structure. We denote by U the set of invertible elements of V.

9.1. Recall that a subspace I of V is said to be an *ideal* of $\mathscr{S}$ if for all $i \in I$ the rational map
$$x \mapsto j(x+i) - jx$$
is a rational map of V in I, see 1.3.

Let G° be the identity component of the structure group G of $\mathscr{S}$. The ideal I is called a *characteristic ideal* if it is a G°-stable subspace of V.

If $\mathscr{S}$ is defined over the subfield k of K then the ideal I is said to be *defined over k* if it is a k-subspace of V.

9.2. Let $I \neq V$ be an ideal of $\mathscr{S}$. Put $V' = V/I$, let $\phi: V \to V'$ be the canonical linear map. Put $e' = \phi e$.
From the definition of the notion of ideal it follows that if $x \in U$ we have
$$\phi(j(y)) = \phi(j(x)),$$
for all $y \in (x+I) \cap U$. It follows that

(1) $$j'(\phi(x)) = \phi(j(x)) \quad (x \in U)$$

defines a rational map $j': V' \to V'$. From axiom (J1) of §1 it follows that $j' \circ j' = \text{id}$, so that j' is in fact birational. It is trivial to check that j' is homogeneous of degree -1 and that $\mathscr{S}' = (V', j', e')$ satisfies axiom (J2) of 1.3.

It then follows that $e' \neq 0$. For if we had $e' = 0$, the formula of (J2) would imply $j' + j' \circ j' = 0$, which contradicts the homogeneity of j'. We then see that $\mathscr{S}'$ satisfies axioms (J1) and (J2) of 1.3.

Let $x \in U$, put $x' = \phi(x)$. Then j' is regular in x'. It follows from (1) that

(2) $$(dj')_{x'} \circ \phi = \phi \circ (dj)_x.$$

Define a linear transformation $P'(x')$ $(x' \in \phi(U))$ of V' by
$$P'(x')^{-1} = -(dj')_{x'}^{-1}.$$

From §1, (16) and (2) it follows that we have, P denoting the quadratic map of $\mathscr{S}$,

(3) $$P'(x') \circ \phi = \phi \circ P(x).$$

3.4 then shows that P' extends to a quadratic map P' of V' into its space of endomorphisms $\operatorname{End}(V')$.

9.3 Proposition. $\mathscr{Q}' = (V', P', e')$ *is a quadratic Jordan algebra.*

(V, P, e) is quadratic Jordan algebra by 7.10(i). We have remarked in 9.2 that $e' \neq 0$. It then follows from (3) that the axioms of 7.1 for a quadratic Jordan algebra hold in $\mathscr{Q}'$.

9.4 Proposition. $\mathscr{S}' = (V', j', e')$ *is a J-structure if one of the following conditions is satisfied:*
(a) $\operatorname{char}(K) \neq 2$;
(b) $\operatorname{char}(K) = 2$ *and* $\mathscr{S}$ *is separable* (7.7);
(c) I *is a characteristic ideal.*

We have seen in 9.2 that $e' \neq 0$ and that (J1) and (J2) hold in $\mathscr{S}'$. So it remains to prove axiom (J3) of 1.3.

Let G' be the structure group of $\mathscr{S}'$. By 1.16(v) we know that $P(x)$ ($x \in U$) is the structure group G of $\mathscr{S}$. It then follows from (1) and (3) that $P'(x')$ $(x' \in \phi(U))$ lies in G'.

Now if (a) or (b) holds, it follows as in the proof of 7.10(ii) that the set of squares of elements of U contains an open subset of V. It then follows that the set of elements $P'(x') e'$ $(x' \in \phi(U))$ contains an open subset of V', showing that $G' \cdot e'$ contains an open subset of V'. This proves (J3) in cases (a) and (b).

Now assume that (c) holds. Let $g \in G^\circ$. The assumption (c) implies that g induces a linear transformation g' of V'. From (1) and the definition of structure groups (see 1.2) it follows that $g' \in G'$. Moreover we have $\phi(g e) = g' \cdot e'$. Since $\mathscr{S}$ is a J-structure, Ge is open in V. By 1.13 it then follows that $G^\circ \cdot e$ is open in V, hence $\phi(G^\circ \cdot e)$ is open in V'. But $\phi(G^\circ \cdot e)$ is contained in $G' \cdot e'$, hence $G' \cdot e'$ is open in V'. This concludes the proof of 9.4.

Remark. It is not known to the author whether $\mathscr{S}'$ is always a J-structure.

9.5. With the previous notations we say that *the quotient of $\mathscr{S}$ by the ideal I exists* if $\mathscr{S}'$ is a J-structure. In that case we also write $\mathscr{S}' = \mathscr{S}/I$. Notice that the quotient always exist if $\operatorname{char}(K) \neq 2$.

We next give a characterization of ideals, using the quadratic map P.

9.6 Proposition. *Let I be a proper subspace of V.*
(i) *I is an ideal of $\mathscr{S}$ if and only if*

§9. Ideals, the Radical

(a) $P(V)I \subset I$, and
(b) $P(I)V \subset I$;
(ii) (b) is a consequence of (a) if $\mathrm{char}(K) \neq 2$.

Let X be an indeterminate over K. If X is in the set U of invertible elements we have a formal power series

$$j(x+yX) = \sum_{n=0}^{\infty} a_n(x, y) X^n \quad (y \in V),$$

where a_n is a rational map $V \times V \to V$ which is polynomial in its second variable (see 0.9). Using 1.16(iii) and §3, (8) we find

(4) $\begin{cases} a_0(x, y) = jx, \\ a_1(x, y) = -P(x)^{-1} y, \\ P(x) a_n(x, y) = -P(x, y) a_{n-1}(x, y) - P(y) a_{n-2}(x, y) & (n \geq 2), \end{cases}$

if $x \in U$, $y \in V$.

We claim that I is an ideal if and only if we have $a_n(x, y) \in I$ for $n \geq 1$, $x \in U$, $y \in I$. For let $(e_i)_{1 \leq i \leq a}$ be a basis for V such that $(e_i)_{1 \leq i \leq b}$ is one of I. Put $a_n(x, y) = \sum_{i=0}^{a} a_{ni}(x, y) e_i$. From the definition of an ideal one concludes that I is an ideal if and only if $a_{ni}(x, y) = 0$ for $n \geq 1$, $b+1 \leq i \leq a$, $x \in U$, $y \in I$, which establishes our claim.

Now suppose that (a) and (b) hold. It then follows from (a) and the second formula (4) that $a_1(x, y) \in I$ ($x \in U$, $y \in I$). From (a) and (b) one then obtains, using the last formula (4), that $a_2(x, y) \in I$ ($x \in U$, $y \in I$). It then follows by induction that $a_n(x, y) \in I$ for $n \geq 1$, $x \in U$, $y \in I$. Hence I is an ideal.

Conversely, suppose that I is an ideal. Then $a_1(x, y) \in I$ for $x \in U$, $y \in I$. By the second formula (4) this gives $P(U)^{-1} I \subset I$. Since $j(U)$ is dense in V, this implies (a). (b) is proved similarly, using the last formula (4) with $n = 2$. This proves (i).

Next let $\mathrm{char}(K) \neq 2$. By §3, (6) and §3, (10) we have

$$P(y) x^2 = P(x)^{-1} P(P(x) y) e = \tfrac{1}{2} P(x)^{-1} P(P(x) y, e) P(x) y,$$

if $x \in U$. Taking $y \in I$ in this formula, one sees that (a) implies that $P(y)x^2 \in I$ for all $x \in U$ and $y \in I$. Since the set of squares of elements of U is dense in V (see the proof of 7.10(ii)), it follows that $P(I)V \subset I$. This proves (ii).

9.7 Corollary. *Let I be a proper subspace of V.*
(i) *I is a characteristic ideal of $\mathscr{S}$ if and only if*
(a) *I is a G°-stable subspace of V and*
(b) *$I^2 \subset I$;*
(ii) *(b) is a consequence of (a) if $\mathrm{char}(K) \neq 2$.*

If I is a characteristic ideal, then (a) is true by definition and (b) follows from 9.6(i)(b). Next assume that (a) and (b) hold. Since $P(x) \in G°$ for $x \in U$ (by 1.16(v)), it follows that (a) implies 9.6(i)(a). By 1.16(i) we have if $g \in G°$, $x \in I$
$$P(x)(ge) = (g')^{-1}(g'x)^2,$$
where $g' \in G°$.

From (a) and (b) it follows that the right-hand side lies in I. Hence $P(I)(G° \cdot e) \in I$. From 1.13 we then conclude that $P(I) V \in I$. Hence 9.6(i)(b) also holds, so that I is an ideal by 9.6(i). It is then of course a characteristic ideal. (ii) is a consequence of 9.6(ii).

9.8. If $\text{char}(K) \neq 2$ one can introduce in V a Jordan algebra structure with the properties of 6.5(ii). Using what was established in §6, it easily follows from 9.6 that I is an ideal of $\mathscr{S}$ if and only if it is an ideal for the Jordan algebra structure.

As in the theory of quadratic Jordan algebras we say that a proper subspace I of V is an *inner ideal* of $\mathscr{S}$ if $P(I) V \subset V$ and an *outer ideal* of $\mathscr{S}$ if $P(V) I \subset I$ (see [15, 1.29]). The outer ideals of $\mathscr{S}$ are the proper subspaces of V which are stable for the inner structure group G_1 (defined in 1.19).

An ideal is then by 9.6(i) a proper subspace of V which is both an inner and an outer ideal. By 9.6(ii) the notions of ideal and outer ideal coincide if $\text{char}(K) \neq 2$. This is not the case in characteristic 2, as the following example shows.

9.9 Example. Let $\text{char}(K) = 2$. Let $\mathscr{S}_r$ be the J-structure of (2.10) with $r > 1$. Its underlying vector space is the space $\mathbb{S}_r$ of symmetric $r \times r$ matrices. From the definition of $\mathscr{S}_r$ it follows, using what was said in 2.2 that the quadratic map of $\mathscr{S}_r$ is given by
$$P(X) Y = XYX \quad (X, Y \in \mathbb{S}_r),$$
where the product is the usual matrix product. Now since $\text{char}(K) = 2$, the space $\mathbb{A}_r$ of alternating $r \times r$ matrices is a subspace of $\mathbb{S}_r$ and it is trivial to verify that $P(\mathbb{S}_r) \mathbb{A}_r \subset \mathbb{A}_r$. Hence $\mathbb{A}_r$ is an outer ideal in $\mathscr{S}_r$. But is not an inner ideal: there exist alternating matrices whose square is not alternating.

9.10 Radical elements, the radical. $r \in V$ is called a *radical element* of $\mathscr{S}$ if for all g in the structure group G we have that gr is nilpotent. The set of all radical elements is called the *radical* of $\mathscr{S}$. $\mathscr{S}$ is called a *semisimple* J-structure if 0 is the only radical element of $\mathscr{S}$. The next result shows that semisimplicity is equivalent to nondegeneracy of the norm N of $\mathscr{S}$ (see 0.15).

9.11 Lemma. $r \in V$ *is a radical element if and only if*
(5) $$N(x+r) = N(x) \quad (x \in V).$$

§9. Ideals, the Radical

Let $g \in G$. By 1.5 there exists $c \in K^*$ such that
$$N(g\,x) = c\,N(x) \quad (x \in V).$$
Putting $x = e$ it follows that $c = N(g\,e)$. Hence

(6) $$N(g\,e + r) = N(g\,e)\,N(e + g^{-1} \cdot r).$$

If r is a radical elements then $g^{-1} \cdot r$ is nilpotent for all $g \in G$. It follows from 3.15(iii) that $N(e + g^{-1} \cdot r) = 1$, hence (5) holds for all $x \in Ge$. Since Ge is open in V, (5) follows.

Conversely, if (5) holds it follows from (6) that $N(e + g\,r) = 1$ for all $g \in G$. Since all nonzero scalar multiplications are in G, we also have that $N(e + t\,g\,r) = 1$ ($t \in K^*$, $g \in G$). By 3.15(iii) we conclude that $g\,r$ is nilpotent for all $g \in G$.

The next result gives the main properties of the radical. We say that an ideal I of $\mathscr{S}$ is a *nil ideal* if all its elements are nilpotent.

9.12 Theorem. *Let R be the radical of $\mathscr{S}$.*
(i) *R is a characteristic ideal of $\mathscr{S}$;*
(ii) *R is the maximal nil ideal of $\mathscr{S}$.*

It follows from 9.11 that the sum of two radical elements of $\mathscr{S}$ is again a radical element. This implies that R is a subspace of V. It is a proper subspace (for $e \notin R$). R is G-stable by the definition of radical elements. To prove that R is a characteristic ideal it suffices, by 9.7(i), to prove that $R^2 \subset R$. By 9.7(ii) this is so if $\mathrm{char}(K) \neq 2$.

So assume now $\mathrm{char}(K) = 2$, let $r \in R$. Assume that $n \geq 2$, $r^{2^n} \in R$. Since r is nilpotent, this is certainly true for large enough n. We shall prove that $r^{2^{n-1}} \in R$. It then follows that $r^2 \in R$, which will finish the proof of (i).

By §3, (14) (with $y = e$) we have $N(x^2) = N(x)^2$ ($x \in V$). Hence
$$N(x + r^{2^{n-1}})^2 = N((x + r^{2^{n-1}})^2) = N(x^2 + r^{2^n} + P(x, r^{2^{n-1}})\,e),$$
(using §3, (19)). Since $r^{2^n} \in R$ it follows from 9.11 and §3, (10) that

(7) $$N(x + r^{2^{n-1}})^2 = N(x^2 + P(x, e)\,r^{2^{n-1}}).$$

From §3, (22) we obtain
$$P(x, e)\,r^{2^{n-1}} = P(x, r)\,r^{2^{n-1}-1} = P(x, r)\,P(r)^{2^{n-2}-1}\,r$$
(using §3, (19)). Since R is G-stable it is an outer ideal of $\mathscr{S}$. But then the last formula shows that $P(x, e)\,r^{2^{n-1}} \in R$. (7) now shows that
$$N(x + r^{2^{n-1}})^2 = N(x^2) = N(x)^2,$$
which implies that $r^{2^{n-1}} \in R$ (via 9.11), as we claimed. This proves (i).

We come now to the proof of (ii). It follows from (i) that R is a nil ideal. Hence to prove (ii) it suffices to show that any nil ideal I is a subspace of R. Let $r \in I$. We then have

$$N(x+r)^2 = N(x^2 + r^2 + P(x,r)e) = N(x^2 + r^2 + P(x,e)r) \quad (x \in V)$$

(by §3, (10)).

Suppose that x is invertible. Using §3, (14) we then can conclude that

$$N(x+r)^2 = N(x^2) N(e + P(x)^{-1} \cdot r^2 + P(x)^{-1} P(x,e) r).$$

Since I is an ideal, we have by 9.6(i) that

$$P(x)^{-1} r^2 + P(x)^{-1} P(x,e) r \in I.$$

Since I is a nil ideal it follows from 3.15(iii) that

$$N(x+r)^2 = N(x^2) = N(x)^2 \quad (x \in U).$$

But then we have $N(x+r) = \varepsilon N(x)$ $(x \in V)$, where $\varepsilon = \pm 1$. Taking $x = e$ it follows by 3.15(iii) that $\varepsilon = 1$. 9.11 now shows that $r \in R$. Hence $I \subset R$, which proves (ii).

9.13 Corollary. (i) *The quotient of $\mathscr{S}$ by its radical R exists;*
(ii) *$\mathscr{S}/R$ is semisimple.*

By 9.12(i) we know that R is a characteristic ideal. (i) is then a consequence of 9.4(c).

Put $\mathscr{S}/R = \mathscr{S}' = (V', j', e')$. Let ϕ be the canonical homomorphism $V \to V' = V/R$. We have $j' \circ \phi = \phi \circ j$. The definition of powers (see 3.10) then shows that $\phi(x^n) = \phi(x)^n$ $(x \in V)$. It follows that if $\phi(x)$ is nilpotent some power of x lies in R. Since R is a nil ideal by 9.12(ii), it follows that x is nilpotent.

Let $g \in G$ (the structure group of $\mathscr{S}$). Then g induces a linear transformation g' of V', which lies in the structure group of $\mathscr{S}'$. Suppose now that $\phi(x)$ is a radical element of S'. Then $\phi(gx)$ is nilpotent for all $g \in G$. By what we said before, this implies that gx is nilpotent $(g \in G)$, so that x is a radical element. Hence $\phi(x) = 0$. This proves (ii).

We terminate this section with some results on ideals which will be needed later.

9.14 Proposition. *Let I be an ideal in $\mathscr{S}$. Let $x \in I$, suppose that $x = x_s + x_n$ is the Jordan decomposition of x (see 8.9). Then $x_s, x_n \in I$. Moreover x_s is a linear combination of idempotent elements which are contained in I.*

It follows from 9.6(i) and §3, (18) that all powers x^i $(i \geq 1)$ are in I. The first assertion now follows from 8.9. We also see that the substructure $\mathscr{S}(x_s)$ generated by x_s (see 8.8) is contained in I. The last point follows from the fact that $\mathscr{S}(x_s)$ is isomorphic to a direct sum of copies of the 1-dimensional J-structure $\mathscr{J}(K)$.

§9. Ideals, the Radical

9.15 Corollary. *If I is not a nil ideal then I contains an idempotent element $\neq 0$.*

9.16 Proposition. *Suppose that e is the only nonzero idempotent element in V. Then the radical R of $\mathscr{S}$ is a hyperplane in V.*

Let $x \in V$, let $x = x_s + x_n$ be its Jordan decomposition. The hypothesis implies that $\mathscr{S}(x_s) = Ke$. It then follows from 3.15(iii) that the polynomial function $t \mapsto N(te - x)$ ($t \in K$) is a power of a linear polynomial function, for all $x \in V$. From 0.13, using axiom (J3) and 1.5 we see that this implies that N is a power of a linear polynomial. The assertion now follows from 9.11.

9.17 Corollary. *Let $\mathscr{S}$ be a semisimple J-structure such that e is the only nonzero idempotent in V. Then $\mathscr{S}$ is isomorphic to the 1-dimensional J-structure $\mathscr{S}(K)$.*

Notes

In the situation of 9.6, let $\mathscr{Q}$ be the quadratic Jordan algebra defined by $\mathscr{S}$ according to 7.10(i). 9.6 then shows that I is an ideal in $\mathscr{S}$ if and only if it is an ideal in $\mathscr{Q}$, in the sense of [15, 1.29].

Our method of introducing the radical of a J-structure is not a direct transcription of one of the current methods for the radical of a Jordan algebra or a quadratic Jordan algebra, given for example in [8, I, §7], [14, Chapter V], [15, p. 3.5]. However if $\text{char}(K) \neq 2$ our radical coincides with the radical of the Jordan algebra defined by $\mathscr{S}$ according to 6.5. This follows from 9.12(ii), using the results of [14, Ch. V]. By the remarks made in [14, p. 434] it then also follows that our radical coincides with that of [8]. It is likely that if $\text{char}(K) = 2$ the radical introduced here coincides with the radical of the quadratic Jordan algebra $\mathscr{Q}$ (introduced in [22], see also [15, p. 3.5]).

9.16 is an analogue of a well-known result for Jordan algebras, see [14, Theorem 5, p. 198].

§10. Peirce Decomposition Defined by an Idempotent Element

In this section some properties of algebraic tori will be used, which can be found in [1].

10.1. $\mathscr{S} = (V, j, e)$ denotes a J-structure. a is a fixed idempotent element in V (i.e. $a^2 = a$). P is the quadratic map of $\mathscr{S}$, G and G_1 denote the structure group and inner structure group of $\mathscr{S}$, respectively (see §1). Sometimes we shall assume that $\mathscr{S}$ is defined over a subfield k of K. In that case we denote by $\bar{k}$ an algebraic closure of k, contained in K and by k_s the separable closure of k in $\bar{k}$. We then denote (in conformity with the notation of [1, p. 48]) by $V(k)$ the set of k-rational points of V.

By 3.14 we know that for all $t, u \in K^*$ the elements $t a + u(e-a)$ of V are invertible. P denoting the quadratic map of $\mathscr{S}$, it follows that

$$\phi_a(t, u) = P(t a + u(e-a))$$

defines a morphism (of algebraic varieties) of the 2-dimensional torus $(\mathbb{GL}_1)^2$ into G_1.

10.2 Lemma. (i) ϕ_a *is a homomorphism of algebraic groups. If* $a \neq 0, e$ *the image of* ϕ_a *is a 2-dimensional subtorus* S_a *of* G_1, *hence also of* G;
(ii) *If* S *is defined over* k *and if* $a \in V(k)$, *then* ϕ_a *and* S_a *are defined over* k *and* S_a *is a* k-*split subtorus of* G_1.

According to 3.14 we have

$$j(t a + u(e-a)) = t^{-1} a + u^{-1}(e-a) \quad (t, u \in K^*),$$

whence by differentiation

$$(dj)_{t a + u(e-a)}(t_1 a + u_1(e-a)) = -t^{-2} t_1 a - u^{-2} u_1(e-a).$$

By §1, (16) it follows that

(1) $\qquad P(t a + u(e-a))(t_1 a + u_1(e-a)) = t^2 t_1 a + u^2 u_1(e-a).$

By continuity this holds for all $t, u, t_1, u_1 \in K$. From §3, (13) we see that all $\phi_a(t, u)$ $(t, u \in K^*)$ commute.

§10. Peirce Decomposition Defined by an Idempotent Element

We then have by §3, (6) and (1)

$$\phi_a(t^2 t_1, u^2 u_1) = \phi_a(t, u) \phi_a(t_1, u_1) \phi_a(t, u) = \phi_a(t, u)^2 \phi_a(t_1, u_1).$$

Taking $t_1 = u_1 = 1$ it follows that $\phi_a(t^2, u^2) = \phi_a(t, u)^2$. Hence

$$\phi_a(t^2 t_1, u^2 u_1) = \phi_a(t^2, u^2) \phi_a(t_1, u_1),$$

which implies that ϕ_a is a homomorphism.

Next let $a \neq 0, e$. Clearly $\dim \phi_a((\mathbb{GL}_1)^2) \leq 2$. Since $\phi_a(t, t)$ is scalar multiplication by t^2, it follows that $\dim S_a \geq 1$. To prove $\dim S_a = 2$, it therefore suffices to show that S_a does not consist of scalar multiplications only. Suppose the contrary. Since $P(t\, a + u(e - a))$ is a scalar multiplication for all $t, u \in K^*$, it must also be so for $t = 1$, $u = 0$. Hence $P(a)$ is a scalar multiplication. But by §3, (19) we know that $P(a)$ is idempotent, hence $P(a) = \mathrm{id}$ or $P(a) = 0$. Since $P(a) = a$, this violates the assumption that $a \neq 0, e$. This proves (i).

If $\mathcal{S}$ is defined over $k \subset K$, then P is defined over k, from which the assertions of (ii) immediately follow (for the notion of k-split torus see [1, p. 200 and p. 205]).

Define subspaces V_a and V'_a of V by

$$V_a = \{x \in V \mid \phi_a(t, u)\, x = t^2\, x\},$$
$$V'_a = \{x \in V \mid \phi_a(t, u)\, x = t\, u\, x\}.$$

Since $e - a$ is also idempotent (by 3.14, for example) the space V_{e-a} is defined and

$$V_{e-a} = \{x \in V \mid \phi_a(t, u)\, x = u^2\, x\}.$$

We have $V'_{e-a} = V'_a$.

10.3 Proposition. (i) V is the direct sum of the subspaces V_a, V'_a, V_{e-a};
(ii) $V_a = P(a)\, V$, $(P(a, e) - 2)\, V_a = (P(a, e) - 1)\, V'_a = P(a, e)\, V_{e-a} = (P(a) - 1)\, V_a = 0$;
(iii) $P(V_a)\, V'_a = P(V_a)\, V_{e-a} = P(V_a, V'_a)\, V_{e-a} = 0$ and $P(V_a, V_{e-a})\, V_a \subset V_{e-a}$, $P(V_a, V'_a)\, V_a \subset V'_a$, $P(V_a, V'_a)\, V_a \subset V'_a$, $P(V'_a)\, V_a \subset V_a$, $P(V'_a)\, V_a \subset V_{e-a}$;
(iv) If $\mathcal{S}$ is defined over k and $a \in V(k)$, then the subspaces V_a, V'_a, V_{e-a} are defined over k.

ϕ_a defines a rational representation of $(\mathbb{GL}_1)^2$ in V. Let c be a character of $(\mathbb{GL}_1)^2$ (in the sense of [1, p. 164]). The weight space corresponding to c of [loc. cit., p. 166] is then the space of all $x \in V$ such that

$$t^2 P(a)\, x + t\, u\, P(a, e - a)\, x + u^2\, P(e - a)\, x = c(t, u)\, x.$$

The linear independence of characters (see [1, 8.1, p. 199]) implies that we must have $c(t, u) = t^2$, tu or u^2. (i) then follows from the complete reducibility theorem for rational representations of an algebraic torus (see [loc. cit., p. 204]).

It also follows that if $x \in V_a$ we have

$$P(a) x = a, \quad P(a, e-a) x = P(e-a) x = 0,$$

and that if $x \in V'_a$ we have

$$P(a, e-a) x = x, \quad P(a) x = P(e-a) x = 0.$$

Hence $P(a) V = P(a)(V_a + V'_a + V_{e-a}) = P(a) V_a = V_a$. Also, $(P(a, e) - 2) V_a = P(a, e-a) V_a = 0$. From the second set of formulas it follows that $(P(a, e) - 1) V'_a = 0$. Also $P(a, e) V_{e-a} = (P(e-a, e) - 2) V_{e-a} = 0$. This establishes (ii).

By §3, (6) we have

$$\phi_a(t, u) P(x) y = P(\phi_a(t, u) x) \phi_a(t, u)^{-1} y.$$

If $x, y \in V_a$, this implies that

$$\phi_a(t, u) P(x) y = t^2 P(x) y,$$

hence $P(V_a) V_a \subset V_a$. The other statements of (iii) are established in a similar manner. Finally, (iv) follows from 10.2(ii) and [1, p. 200].

The decomposition $V = V_a + V'_a + V_{e-a}$ is called the *Peirce decomposition* of V defined by the idempotent a.

10.4 Examples. (1) Let A be a finite dimensional associative algebra with identity e, let $\mathscr{J}(A)$ be the corresponding J-structure (see 2.1). Let $a \neq 0, e$ be an idempotent element of A. It is also an idempotent in $\mathscr{J}(A)$. Since we now have $P(x) y = x y x$ (see 2.2), it follows that $A_a = a A a$. It is easily seen that $A'_a = a A(e-a) + (e-a) A a$.
(2) Let V be a finite dimensional vector space, let Q be a quadratic form on V and let $e \in V$ be such that $Q(e) = 1$. With the notation of 2.17, suppose that $Q_e \neq 0$. Let $\mathscr{S}$ be the J-structure of 2.17. Let $a \neq 0, e$ be an idempotent. It follows from §2, (13) that we have $Q(a) = 0$, $Q(a, e) = 1$. Moreover, $P(a) x = Q(a, \bar{x}) a$. A simple computation shows that

$$V_a = K a, \quad V_{e-a} = K(e-a), \quad V'_a = \{x \in V \mid Q(e, x) = Q(a, x) = 0\}.$$

10.5 Proposition. *Let $a \neq 0, e$ be an idempotent. There exists a unique birational transformation j_a of V_a with the following properties.*
(i) *If $x \in V_a$, $y \in V_{e-a}$ are such that $x + y$ is invertible, then j_a is regular in x and*

(2) $$j_a(x) = P(a) j(x+y),$$

§ 10. Peirce Decomposition Defined by an Idempotent Element

We then have

(3) $$j(x+y) = j_a(x) + j_{e-a}(y);$$

(ii) $\mathscr{S}_a = (V_a, j_a, a)$ *satisfies the axioms* (J1), (J2) *of* 1.3;
(iii) *If* $\mathscr{S}$ *is defined over* $k \subset K$ *and* $a \in V(k)$, *then* j_a *is defined over* k.

Let N be the norm of $\mathscr{S}$. The set U of the $x \in V_a$ such that there exists $y \in V_{e-a}$ with $x+y$ invertible is the complement of the closed set consisting of the $x \in V_a$ with $N(x+y) = 0$ for all $y \in V_{e-a}$. Hence U is open. U is nonempty because $a \in U$. The uniqueness of j_a satisfying (2) is then clear.

Let x and y be as in (i) and put

$$j(x+y) = v + v' + w \quad (v \in V_a, v' \in V'_a, w \in V_{e-a}).$$

From the definition of V_a and V'_a it follows, using 1.16(v) and 10.3 that

(4) $$j(t^{-2}x + u^{-2}y) = t^2 v + tu v' + u^2 w \quad (t, u \in K^*).$$

$(t, u) \mapsto j(t^{-2}x + u^{-2}y)$ defines a rational map $K \times K \to V$, whose denominator and numerator only involve powers $t^{2i} u^{2j}$. But it then follows that the right-hand side of (4) can only involve even powers of t and u, so that we must have $v' = 0$.

Define a rational map ψ of $V_a \times V_{e-a}$ into V_a by

$$\psi(x, y) = P(a) j(x+y),$$

for $x+y$ invertible. From (4) it then follows that

$$\psi(x, y) = \psi(x, ty) \quad (t \in K^*).$$

But this implies that $\psi(x, y)$ is independent of y. This proves the existence of a rational map j_a satisfying (2).

Since

$$j(x+y) = P(a) j(x+y) + P(e-a) j(x+y),$$

if x and y are as in (i), it also follows that (3) holds. The axioms (J1) and (J2) for $\mathscr{S}$ imply that $\mathscr{S}_a$ satisfies these axioms, too. Hence (ii) holds. In particular, j_a is birational. The proof of (i) is now also complete. The last point is clear from (2).

10.6 Corollary. *We have* $P(V_a, V_{e-a}) V_a = P(V_a, V_{e-a}) V_{e-a} = 0$.

It follows from (3) by differentiation, using §1, (16), that

$$-P(x+y)^{-1}(v+w) = (dj_a)_x(v) + (dj_{e-a})_y(w),$$

if x and y are as in 10.5(i) and if $v \in V_a$, $w \in V_{e-a}$. Using 10.3(iii) this implies that

(5) $$P(x)(dj_a)_x(v) + P(x,y)(dj_{e-a})_y(w) = -v.$$

Taking $v=0$ in (5) we get

$$P(x,y)(dj_{e-a})_y(w) = 0,$$

from which it follows that $P(x,y)w=0$ for x and y in suitable nonempty open subsets of V_a and V_{e-a}, and for all $w \in V_{e-a}$. This implies $P(V_a, V_{e-a})V_{e-a} = 0$. Replacing a by $e-a$ we obtain $P(V_a, V_{e-a})V_a = 0$.

10.7 Corollary. (i) *There exists a quadratic map $P_a: V_a \to \mathrm{End}(V_a)$ such that $P_a(x) = -(dj_a)_x^{-1}$ if j_a is regular in $x \in V_a$;*
(ii) *We have $P(x)y = P_a(x)y$ $(x, y \in V_a)$;*
(iii) *If j_a is regular in $x \in V_a$, we have $P_a(x) \circ j_a = j_a \circ P_a(x)^{-1}$, in particular it follows that then $P_a(x)$ lies in the structure group of j_a;*
(iv) *(V_a, P_a, a) is a quadratic Jordan algebra; if $\mathrm{char}(K) \neq 2$ we have that $\mathscr{S}_a = (V_a, j_a, a)$ is a J-structure.*

Putting $P_a(x)y = P(x)y$ $(x, y \in V_a)$, it follows from 10.3(iii) that P_a is a quadratic map of V_a into $\mathrm{End}(V_a)$.

(5), with $w=0$, shows that P_a has the property asserted in (i). This establishes (i) and (ii). The property of (iii) now follows from the corresponding one of P (see 1.16(v)).

7.10(i) implies that (V_a, P_a, a) is a quadratic Jordan algebra, using (ii). The last assertion of (iv) is then a consequence of 7.10(ii), taking into account that j_a is the inversion of the quadratic Jordan algebra (V_a, P_a, a) (which follows from (ii)).

We next discuss a particular case in which $\mathscr{S}_a$ is always a J-structure.

10.8 Proposition. *Suppose $a \neq 0$, e is an idempotent such that $V_a' = 0$. Then V_a and V_{e-a} are characteristic ideals of $\mathscr{S}$. $\mathscr{S}_a$ and $\mathscr{S}_{e-a}$ are J-structures and $\mathscr{S}$ is the direct sum $\mathscr{S}_a \oplus \mathscr{S}_{e-a}$.*

From 9.6(i), 10.3(i) and 10.6 it follows that V_a is an ideal. To prove that it is a characteristic ideal we have to show that it is G°-stable (see 9.1). Let $g \in G^\circ$, let g' be the image of g under the standard automorphism of G. If $x \in V_a$, $y \in V_{e-a}$ are such that $x+y$ is invertible, we then have as a consequence of (2)

(6) $$j_a(P(a)g' \cdot x + tP(a)g' \cdot y) = P(a)(g(j_a x) + t^{-1}g(j_{e-a}y)),$$

for all $t \in K^*$.

There is a nonempty open subset U of G°, containing the neutral element, such that $x \mapsto P(a)g'x$ is a nonsingular linear transformation of V_a if $g \in U$. Then the left-hand side of (6), considered as a function of t,

§ 10. Peirce Decomposition Defined by an Idempotent Element   95

is regular for $t=0$. This implies that we must have $P(a)\,g(j_{e-a}\,y)=0$, which shows that $g\,V_{e-a}\subset V_{e-a}$ for $g\in U$. It follows that V_{e-a} is G°-stable. The same is true for V_a.

It now follows from 1.13 that the orbit $G^\circ\cdot a$ is open in V_a. Since the restrictions to V_a of the transformations of G° lie in the structure group of j_a (as follows from (2)), we conclude that $\mathscr{S}_a$ satisfies axiom (J 3). 10.5 (ii) now implies that $\mathscr{S}_a$ is a J-structure. The same is true for $\mathscr{S}_{e-a}$. It follows from (3) that $\mathscr{S}=\mathscr{S}_a\oplus\mathscr{S}_{e-a}$ (see 1.25).

If a has the property $V_a'=0$ of 10.8 we say that a is a *central idempotent*. The following proposition gives a rationality result on central idempotents.

10.9 Proposition. *Suppose that $\mathscr{S}$ is defined over k. Let $a\in V(\bar{k})$ be a central idempotent. Then $a\in V(k_s)$.*

Let l and m be subfields of K with $k\subset l\subset m$. Let D be an l-derivation of m. Since $V(m)$ can be identified with $m\otimes_l V(l)$, the derivation D extends to an l-linear transformation of $V(m)$, also denoted by D, which is such that

$$D(t\,x)=(D\,t)\,x+t\,D\,x \quad (x\in V(m),\ t\in m).$$

It is an easy matter to check (for example by expressing everything in terms of a basis of $V(l)$) that we have

(7) $\qquad D\bigl(P(x)\,y\bigr)=P(x,D\,x)\,y+P(x)\,D\,y \quad (x,y\in V(m)).$

We now establish the following lemma.

10.10 Lemma. *Let $a\in V(m)$ be a central idempotent. Then $D\,a=0$.*

Apply (7) with $x=a$, $y=e$. We obtain that $D\,a=P(a,D\,a)\,e=P(a,e)\,D\,a$. By 10.3 (ii) this shows that $D\,a\in V_a'=0$.

We can now prove 10.9. If $a\in V(\bar{k})$ is a central idempotent, there is a finite extension $m\subset\bar{k}$ of k such that $a\in V(m)$. We may assume that $\operatorname{char}(K)=p>0$. Let l be a subfield of m, containing k, such that m/l is purely inseparable.

We claim that $a\in V(l)$. It suffices to prove this in the case that m has degree p over l. But in that case it is easily seen that there exists an l-derivation D with $\operatorname{Ker} D=l$. Our claim then follows from 10.10. We conclude that $a\in V(m\cap k_s)$, which establishes 10.9.

10.11. Let N be the norm of $\mathscr{S}$, let d be its degree. We define a polynomial function N_a on V_a by

$$N_a(x)=N(x+e-a) \quad (x\in V_a).$$

If $\mathscr{S}$ is defined over $k\in K$ and $a\in V(k)$, then N_a is defined over k. We assume from now on that $a\neq 0, e$.

10.12 Lemma. (i) N_a *is a homogeneous polynomial function. We have* $N_a(a)=1$ *and*

(8) $$N(x+y)=N_a(x)N_{e-a}(y) \quad (x\in V_a, y\in V_{e-a});$$

(ii) *Let* Φ_i *be as in* 1.11. *Then*

(9) $$N_a(x)N_a(j_a x+y)=\sum_{i=0}^{d}\Phi_i(x,y),$$

if j_a *is regular in* $x\in V_a$, *for all* $y\in V_a$. *In particular we have* $N_a(x)N_a(j_a x)=1$;
(iii) *If* V_a+V_{e-a} *contains an element whose minimum polynomial has the same degree as* N, *then* N_a *is a denominator of* j_a.

Let $x\in V_a$, $y\in V_{e-a}$ be such that $x+y$ is invertible. It follows from (3), using 1.8, that the polynomial function $(x,y)\mapsto N(x+y)$ on $V_a\times V_{e-a}$ divides a power of the product of the denominators of j_a and j_{e-a}. This implies that there exist homogeneous polynomial functions N_a' and N_{e-a}' on V_a and V_{e-a}, respectively, such that

$$N(x+y)=N_a'(x)N_{e-a}'(y).$$

We may normalize such that $N_a'(a)=N_{e-a}'(e-a)=1$. Putting $y=e-a$ it follows that $N_a=N_a'$. This proves (i).

Let $x, y\in V_a$, $v, w\in V_{e-a}$ and suppose that $x+v$ is invertible. Put

$$\Phi_a(x,y)=N_a(x)N_a(j_a(x)+y),$$
$$\Phi_{e-a}(v,w)=N_{e-a}(v)N_{e-a}(j_{e-a}(v)+w).$$

From what was established in 1.11 we conclude, using (3) and (8), that

$$\Phi_a(x,y)\Phi_{e-a}(v,w)=\sum_{i=0}^{d}\Phi_i(x+v, y+w).$$

The right-hand side is a polynomial function in its variables x, y, v, w, which is nonzero for $x=y=0$, $v=w=0$ (by 1.14). Hence Φ_a and Φ_{e-a} must be polynomial functions, which are nonzero for $x=y=0$, $v=w=0$, respectively. Putting $v=w=0$ we obtain (9) up to a nonzero constant. This constant is 1, as is seen by taking $x=a$, $y=0$. The last statement of (ii) is (9) with $y=0$. (iii) readily follows from (3) and (8), taking into account the definition of minimum polynomials (see § 8).

10.13 Lemma. *Let* Φ_i *be as in* 1.11. *If* $x, y\in V_a$, $x'\in V_a$, $z\in V_{e-a}$ *we have*

(10) $$\Phi_i(x+x'+z, y)=\Phi_i(x, y).$$

It follows from the results of 1.11 and 1.16(v) that, ϕ_a being as in 10.1, we have

$$\Phi_i(\phi_a(t,u)v, w)=\Phi_i(v, \phi_a(t,u)w),$$

§ 10. Peirce Decomposition Defined by an Idempotent Element

for all $v, w \in V$. This implies
$$\Phi_i(t^2 x + t u x' + u^2 z, y) = \Phi_i(x + x' + z, t^2 y)$$
for $t, u \in K^*$. By continuity this holds for all $t, u \in K$. Putting $t = 1, u = 0$ the assertion follows.

10.14 Proposition. *Let $r \in V_a$ be such that $N_a(x+r) = N_a(x)$ for all $x \in V_a$. Then r is in the radical of $\mathcal{S}$.*

10.12(ii) implies that $\Phi_i(x, r) = 0$ for $1 \leq i \leq d$, $x \in V_a$. It then follows from (10) that $\Phi_i(v, r) = 0$ for $1 \leq i \leq d$ and all $v \in V$.

The definition of Φ_i (see 1.11) shows that
$$N(v) N(j v + r) = 1$$
for all invertible $v \in V$. From 1.8 and 9.11 we then conclude that r is in the radical of $\mathcal{S}$.

10.15 Corollary. *Assume that $\mathcal{S}$ is semisimple and that $\mathcal{S}_a = (V_a, j_a, a)$ is a J-structure. Then $\mathcal{S}_a$ is semisimple.*

It follows from (3), using what was remarked at the beginning of the proof of 10.12, that the denominator of j_a has the same irreducible factors as N_a. The assertion then follows from 9.11. Notice that $\mathcal{S}_a$ is certainly a J-structure if $\mathrm{char}(K) \neq 2$ (see 10.7(iv)).

We next discuss orthogonality of idempotents and the properties of primitive idempotents.

10.16 Lemma. *Let a and b be idempotent in V (not 0 or e).*
(i) *If $\mathrm{char}(K) \neq 2$ then $a + b$ is idempotent if and only if $b \in V_{e-a}$;*
(ii) *If $b \in V_{e-a}$ then $a \in V_{e-b}$. In that case $a + b$ is idempotent and we have $V_a \subset V_{a+b}$.*

Since $(a+b)^2 = a + b + P(a, b) e$, we see that $a + b$ is idempotent if and only if $P(a, b) e = P(a, e) b = 0$. If $\mathrm{char}(K) \neq 2$ it follows from 10.3(ii) that $P(a, e) b = 0$ if and only if $b \in V_{e-a}$. This proves (i).

If $b \in V_{e-a}$ then 10.6 implies that $P(a, b) e = P(a, b) a + P(a, b)(e-a) = 0$. Hence $a + b$ is idempotent. We have $P(b) a = 0$ by 10.3(iii), so $a \in V_{e-b}$ by 10.3(ii). Finally $P(a, b) V_a = P(b) V_a = 0$ (by 10.6 and 10.3(iii)). Hence $V_a = P(a+b) V_a \in P(a+b) V = V_{a+b}$.

If a and b are as in 10.16(ii) we say that they are *orthogonal idempotents*. 10.16(ii) shows that orthogonality is a symmetric relation between idempotents. We say that $a \neq 0$ is a *primitive idempotent* if a is the only idempotent of $\mathcal{S}$ contained in V_a.

10.17 Lemma. (i) *If a is a primitive idempotent then a is not expressible as a sum of two orthogonal idempotents;*

(ii) *Let I be an ideal in $\mathscr{S}$. Let $a \in I$ be a nonzero idempotent which is not expressible as a sum of two orthogonal idempotents contained in I. Then a is a primitive idempotent.*

Suppose a is a primitive idempotent. Assume $a = b + c$, where b and c are orthogonal idempotents. Since $b \in V_b$ and $V_b \subset V_a$ (by 10.16(ii)), it follows that b is an idempotent in V_a, hence $b = a$. This is a contradiction, whence (i).

Let I be an ideal, let $a \in I$ be a non-primitive idempotent in I. There exists an idempotent $b \neq a$ in V_a. Since $V_a = P(a) V \subset I$ by 9.6(i), we have $b \in I$. By 10.3(ii) we have $P(a,b) e = P(a,e) b = 2b$. Hence $(a-b)^2 = a - b$, so that $a - b$ is idempotent. It also follows that

$$P(e, b)(a-b) = P(a, b) e - P(b, b) e = 0.$$

Since $P(b)(e - (e - a)) = b^2 = P(b) b$, we see that

$$P(e-b)(a-b) = a - b - P(e, b)(a-b) + P(b)(a-b) = a - b,$$

whence $a - b \in V_{e-b}$. Consequently, b and $a - b$ are orthogonal idempotents in I. This proves (ii).

We now prove rationality results about idempotents.

10.18 Lemma. *Suppose that $\mathscr{S}$ is defined over k. Let I be an ideal in $\mathscr{S}$ which is defined over k and which is not a nil ideal.*
(i) *$I(k_s)$ contains a nonzero idempotent;*
(ii) *Let $a \in I(k_s)$ be a nonzero idempotent which is not expressible as a sum of two orthogonal idempotents contained in $I(k_s)$. Then a is a primitive idempotent.*

The set S of idempotents in I is a closed subset of I, which is defined over $\bar{k}$. By 9.15 it is nonempty. Hence, as a consequence of Hilbert's Nullstellensatz, $S \cap I(\bar{k}) \neq \emptyset$. This proves (i) if char$(K) = 0$.

Next let char$(K) = p > 0$. From 3.15(iii) it follows that the set N of nilpotent elements in I is a closed subset. Since I is not nil, $I - N$ is a nonempty open subset of I, which contains an element of $I(k_s)$, e.g. by [1, 13.3, p. 52]. This means that there is a non-nilpotent element x in $I(k_s)$. Let $x = x_s + x_n$ be its Jordan decomposition according to 8.9. Then $x^p = x_s^p + x_n^p$ (this follows from 8.7 and 8.9 if $p \neq 2$ and from 8.9 and 3.11(ii) if $p = 2$). q being a sufficiently large p-power, it follows that $y = x^q = x_s^q$ is a nonzero semisimple element of $I(k_s)$.

Then the substructure $\mathscr{T}$ of $\mathscr{S}$ generated by y is defined (see 8.8), moreover I determines an ideal I' in $\mathscr{T}$, which is not a nil ideal. But $\mathscr{T}$ is isomorphic to a J-structure $\mathscr{J}(A)$, where A is a commutative, associative algebra, which is defined over k_s. In such a J-structure $\mathscr{J}(A)$, any idempotent is central, as follows from 10.4(1). Since $I(\bar{k})$ contains a nonzero

§ 10. Peirce Decomposition Defined by an Idempotent Element

idempotent by what we established at the beginning of the proof, it follows from 10.9 that $I'(k_s)$ contains a nonzero idempotent. Hence $I(k_s)$ does. This establishes (i).

To prove (ii) one argues as in the proof of 10.17(ii). One has to choose the idempotent b occurring in the proof in $V_a(k_s)$, which fact can be established by an argument similar to that used in the proof of (i).

10.19 Proposition. *Suppose $\mathcal{S}$ is defined over k. Let I be an ideal which is defined over k. Any nonzero idempotent $a \in I(k_s)$ is expressible as a sum of mutually orthogonal primitive idempotents of $I(k_s)$. If I is not a nil ideal, then $I(k_s)$ contains primitive idempotents.*

This is a direct consequence of 10.18.

10.20 Proposition. *Let $a \neq 0$, e be a primitive idempotent in $\mathcal{S}$.*
(i) *The set of nilpotents of V_a is a hyperplane in V_a;*
(ii) *Any nilpotent element of V_a lies in the radical of $\mathcal{S}$.*

We proceed as in the proof of 9.16. Let $x \in V_a$, $x = x_s + x_n$ be its Jordan decomposition. Since the powers x^i ($i \geq 1$) lie in V_a, it follows from 8.9 that $x_s, x_n \in V_a$. It follows as in the proof of 9.14 that x_s is a linear combination of idempotents in V_a, whence $x_s \in K a$.

If $\mathcal{S}_a = (V_a, j_a, a)$ is a J-structure one can argue as in the proof of 9.16 to establish (i). By 10.7(iv) this proves (i) if $\text{char}(K) \neq 2$.

Next assume $\text{char}(K) = 2$. We introduce in V_a the restricted Lie algebra structure of 7.15 with $[x, y] = P_a(x, y) a$. Let $\text{ad} x(y) = [x, y]$. If $x \in V_a$ we have that $x^{2^n} \in K a$, for sufficiently large n. It follows that then $(\text{ad} x)^{2^n} = 0$. By Engel's theorem [26, Cor. 2, p. 12] we conclude that the Lie algebra is nilpotent. But then $(x + y)^{2^n} = x^{2^n} + y^{2^n}$, for n large. This implies that the sum of two nilpotent elements of V_a is again nilpotent, from which (i) follows. (ii) follows from (i) by using 10.14.

10.21 Corollary. *Let $\mathcal{S}$ be semisimple, let $a \neq e$ be a primitive idempotent in V. Then $\dim V_a = 1$.*

We can now deal with the basic structure theorem for semisimple J-structures.

10.22 Theorem. *Assume that $\mathcal{S}$ is semisimple. Then V is the direct sum of its nonzero minimal ideals. For any such minimal ideal I there exists a central idempotent $a \in V$ such that $I = V_a$ and that $\mathcal{S}_a = (V_a, j_a, a)$ is a simple J-structure. $\mathcal{S}$ is isomorphic to a direct sum of simple J-structures, which are uniquely determined up to isomorphism (except for their order).*

If $\mathcal{S}$ is simple this is clear (with $V_a = V$). We shall prove that for any proper ideal $I \neq 0$ there exists a central idempotent a with $I = V_a$. It then follows from 10.8 that $\mathcal{S}_a$ is a J-structure. By 10.15 it is semisimple. An induction on dim V now proves the first assertion. The last statement is then easily proved.

So let $I \neq 0$ be a proper ideal. By 9.12 and 9.15, I contains nonzero idempotents. Let $a \in I$ be an idempotent such that dim V_a is maximal. We have $V_a = P(a) V \subset I$ and $V_a' \subset P(a, e) V = P(e, V) a \subset I$ (by 10.3). We shall prove that $V_a = I$, $V_a' = 0$. We proceed in two steps.

(a) $$V_{e-a} \cap I = 0.$$

Let $b \in V_{e-a} \cap I$ be idempotent. By 10.16(ii) we know that $a + b$ is idempotent and that $V_a \subset V_{a+b}$. The maximality of V_a implies that $V_a = V_{a+b}$. Hence $a + b = P(a)(a + b) = a$, whence $b = 0$. This implies that $V_{e-a} \cap I$ consists of nilpotent elements (by 8.9 and 9.14, observing that $V_{e-a} \cap I$ is closed under taking powers with strictly positive exponents).

Let $r \in V_{e-a} \cap I$, let $x \in V_{e-a}$ be such that $a + x$ is invertible. As in the proof of 9.12 we have

$$N_{e-a}(x+r)^2 = N(a+x+r)^2$$
$$= N(a+x)^2 \cdot N(e + P(a+x)^{-1} r^2 + P(a+x)^{-1} P(x, e) r).$$

Let P_{e-a} be as in 10.7. We then have $P(a+x)^{-1} r^2 = P_{e-a}(x)^{-1} r^2 \in V_{e-a} \cap I$ and $P(a+x)^{-1} P(x, e) r = P_{e-a}(x)^{-1} P_{e-a}(x, e-a) r \in V_{e-a} \cap I$. Hence $P(a+x)^{-1} r^2 + P(a+x)^{-1} P(x, e) r$ is nilpotent. It follows from 3.15(iii) that
$$N_{e-a}(x+r)^2 = N(a+x)^2 = N_{e-a}(x)^2,$$

whence (as in the proof of 9.12)

$$N_{e-a}(x+r) = N_{e-a}(x).$$

But now 10.14 shows that $r = 0$. Hence $V_{e-a} \cap I = 0$.

(b) $$V_a' = 0.$$

By 10.3(iii) we have $P(V_a') V_a \subset V_{e-a} \cap I = 0$. Let Φ_i be as in 1.11. It follows from 1.11 that we have, if $x \in V_a$, $y \in V_{e-a}$, $x' \in V_a'$

$$0 = \Phi_i(P(x') x, y) = \Phi_i(x, P(x') y)$$

$(1 \leq i \leq d)$. 10.12(ii) then implies that

$$N_a(x + P(x') y) = N_a(x),$$

which by 10.14 implies that $P(V_a') V_{e-a} = 0$.

Since $(V_a')^2 \subset V_a + V_{e-a}$, it now follows that $P(V_a')(V_a')^2 = 0$. In particular we have $(x')^4 = P(x')(x')^2 = 0$ for all $x' \in V_a'$. So all elements of V_a' are nilpotent.

§ 10. Peirce Decomposition Defined by an Idempotent Element

Let $x \in V_a$, $y \in V_{e-a}$, $x' \in V_a'$. By §3, (14) and 3.15(iii) we have

(11) $$N(P(e+x')(x+y)) = N(x+y).$$

Now

$$P(e+x')(x+y) = x+y+P(x',e)x+P(x',e)y = x+y+t_{x,y}(x')$$

(the other terms vanish by what we just established). We have

$$t_{a,0}(x') = P(x',e)a = P(a,e)x' = x',$$

by 10.3(ii).

By 10.3(iii), $t_{x,y}$ is a linear transformation of V_a', depending linearly on x and y. Since it is surjective for a particular choice of (x,y), it must be surjective for (x,y) in some nonempty open subset U of $V_a \times V_{e-a}$. Then (11) implies

$$N(x+y+x') = N(x+y) \quad ((x,y) \in U).$$

But then we have

$$N(v+x') = N(v)$$

for all $v \in V$. By 9.11 it follows that x' is a radical element, hence $V_a' = 0$. Since $V_a' = V_{e-a} \cap I = 0$, it follows that $I = V_a$, establishing the assertion made in the beginning of the proof.

10.23 Corollary. *Assume that $\mathcal{S}$ is defined over k and semisimple. Then its nonzero minimal ideals are defined over k_s. $\mathcal{S}$ is isomorphic to a direct sum of simple J-structures which are defined over k_s.*

The corollary will follow if we show that the central idempotents of $\mathcal{S}$ are defined over k_s. It follows from 10.8 and 10.22 that the set of central idempotents is finite. On the other hand it is a k-closed subset of the k-open subset $\{x \in V \mid \det(P(x,e)-1) \neq 0\}$ of V (by 10.3(ii)). It follows that all central idempotents lie in $V(\bar{k})$, hence in $V(k_s)$ by 10.9.

10.24. Suppose $\mathcal{S}$ is defined over k. We say that $\mathcal{S}$ is *k-simple*, if $\mathcal{S}$ does not contain ideals I defined over k distinct from 0 and V.

10.22 and 10.23 imply that a semisimple $\mathcal{S}$ is a direct sum of k-simple J-structures. Moreover 10.23 implies that a k-simple $\mathcal{S}$ is a direct sum of simple J-structures, which are defined over k_s and which are permuted transitively by the Galois group of k_s/k.

10.25 Sets of orthogonal idempotents. Let $a_1, \ldots, a_k$ be mutually orthogonal idempotent elements in V, with $a_1 + \cdots + a_k = e$. With the notations of 10.3, 10.5, 10.11 we put

$$V_r = V_{a_r}, \quad j_r = j_{a_r}, \quad N_r = N_{a_r}.$$

10.26 Lemma. (i) *If $x_r \in V_r$ ($1 \leq r \leq h$) is such that $\sum_{r=1}^{h} x_r$ is invertible then*

$$j\left(\sum_{r=1}^{h} x_r\right) = \sum_{r=1}^{h} j_r x_r;$$

(ii) *If $t_r \in K^*$ ($1 \leq r \leq h$) then $\sum_{r=1}^{h} t_r a_r$ is invertible and*

$$j\left(\sum_{r=1}^{h} t_r a_r\right) = \sum_{r=1}^{h} t_r^{-1} a_r;$$

(iii) *The morphism ψ of $(\mathbb{GL}_1)^h$ into the inner structure group G_1 of $\mathscr{S}$ defined by*

$$\psi(t_1, \ldots, t_h) = P\left(\sum_{r=1}^{h} t_r a_r\right)$$

is a homomorphism of algebraic groups;
(iv) *If $\mathscr{S}$ is defined over k and if $a_r \in V(k)$ ($1 \leq r \leq h$) then ψ is defined over k and im ψ is a k-split subtorus of G_1.*

This is true for $h = 2$, by (3), 3.14 and 10.2. Assume the assertion to be true for $h-1$ orthogonal idempotents. Let x_r ($1 \leq r \leq h$) be as in (i). If $r > 1$ we have $x_r \in P(a_r) V \subset P(e - a_1) V = V_{e-a_1}$, by 10.3 and 10.16. Hence by (3)

$$j\left(\sum_{1}^{h} x_r\right) = j_1 x_1 + j_{e-a_1}\left(\sum_{2}^{h} x_r\right).$$

For all $i \neq 1$ we have by 10.16 that $a_1 + a_i$ is idempotent, hence

$$j\left(\sum_{1}^{h} x_r\right) = j_{a_1+a_i}(x_1 + x_i) + \sum_{r \neq 1, i} j_r x_r,$$

by our inductive assumption. These two formulas imply

$$j_{e-a_1}\left(\sum_{2}^{h} x_r\right) = \sum_{2}^{h} j_r x_r,$$

whence (i). (ii) is proved similarly, we omit the proof.

To prove (iii) we first observe that we have, as in the proof of 10.2

(12) $$P\left(\sum_{1}^{h} t_r a_r\right)\left(\sum_{1}^{h} u_s a_s\right) = \sum_{1}^{h} t_r^2 u_r a_r,$$

if $t_r, u_s \in K$. Using §3, (18) it follows that

$$\left(\sum_{1}^{h} t_r a_r\right)^n = \sum_{1}^{h} t_r^n a_r,$$

§ 10. Peirce Decomposition Defined by an Idempotent Element

for all $n \geq 0$. Let $c_1, \ldots, c_h$ be h distinct elements of K^* and put $x = \sum_1^h c_r a_r$.
It is then easily seen that all a_r are linear combinations of powers of x. Using 3.11(ii) it follows that any two of the linear transformations $P(a_r)$, $P(a_r, a_s)$ commute.

Hence all linear transformations $\psi(t_1, \ldots, t_h)$ of (iii) commute. (iii) now follows from (12), in the same manner as the corresponding assertion of 10.2 was derived from (1). The proof of (iv) is easy.

ψ being as in 10.2(iii), put
$$V_{rs} = \{x \in V \mid \psi(t_1, \ldots, t_h) x = t_r t_s x\}.$$
Clearly $V_{rs} = V_{sr}$.

10.27 Proposition. (i) V is the direct sum of the spaces V_{rs} ($1 \leq r \leq s \leq h$);
(ii) We have $V_{rr} = V_r = P(a_r) V$, $V_{rs} = \{x \in V \mid P(a_r, a_s) x = x\}$ ($r \neq s$);
(iii) $P(V_r) V_{rs} = P(V_r) V_s = 0$ ($r \neq s$), $P(V_{rs}) V_t = 0$ if $r \neq s$, $t \neq r$, $t \neq s$;
(iv) If $r \neq s$ then $P(V_{rs}) \sum_{t \neq u} V_{tu} \subset \sum_{t \neq u} V_{tu}$, $P(V_{rs}) V_r \subset V_s$, $(V_{rs})^2 \subset V_r + V_s$.

The proof is quite similar to that of 10.3, we therefore omit it.

10.28 Lemma. Let $x_r \in V_r$ ($1 \leq r \leq h$). Then
$$N\left(\sum_1^h x_r\right) = \prod_1^h N_r(x_r).$$

For $h = 2$ this is (8). The proof for arbitrary h is similar, using 10.26(i) instead of (3).

10.29 Proposition. Let $\mathscr{S}$ be semisimple. Let $a_1, \ldots, a_h$ be mutually orthogonal primitive idempotent elements, with sum e. Assume that there exists an element of the form $\sum_{i=1}^h x_i a_i$ ($x_i \in K^*$) whose minimum polynomial has the same degree d as the norm N. Then $h = d$.

By 10.21 we have dim $V_r = 1$. Clearly (V_r, j_r, e_r) is a J-structure, viz. the 1-dimensional one $\mathscr{J}(K)$. By 10.12(iii) we then have $N_r(t a_r) = t$. 10.28 implies that
$$N\left(\sum_1^h t_r a_r\right) = \prod_1^h t_r,$$
which proves the assertion.

10.30 The equations $P(a, e) = 0$ **and** $P(a) = \mathrm{id}$. If $a \in V$, $P(a, e) = 0$ then we have by § 3, (10)
$$2a = P(e, e) a = P(a, e) e = 0$$
whence $a = 0$ if $\mathrm{char}(K) \neq 2$. This is no longer true if $\mathrm{char}(K) = 2$. For example, let then A be a commutative, associative algebra, let $\mathscr{J}(A)$ be

the J-structure defined by A, see 2.1. It follows from 2.2 that now $P(A, e) = 0$.

From 10.3(ii) one sees that $P(a, e) = 0$ (in characteristic 2) if a is a linear combination of central idempotents. The next result gives a converse to that statement.

10.31 Proposition. *Let* $\operatorname{char}(K) = 2$. *Assume that $\mathscr{S}$ is a semisimple J-structure which is separable. If $a \in V$, $P(a, e) = 0$ then a is a linear combination of central idempotents of $\mathscr{S}$.*

(Separability was defined in 7.7.) Let $C = \{x \in V | P(x, e) = 0\}$. From §3, (7) with $y = z = e$ it follows that

$$P(x, e)^2 = P(x^2, e),$$

hence C is closed under taking squares. It then follows from the last assertion of 8.9 that if $a \in C$ has the Jordan decomposition $a = a_s + a_n$, we have $a_s, a_n \in C$.

We claim that under the assumptions of 10.31, C does not contain nonzero nilpotents. Since the idempotents in C are clearly central, the assertion then follows (the fact that C is closed under taking squares implies that if $a \in C$ is semisimple, the underlying vector space of the substructure $\mathscr{S}(a)$ of 8.8 is contained in C).

To establish the claim let $a \in C$, $a^2 = 0$. Put $y = e$, $z = a$ in §3, (7) and apply both sides to e. We then find that $P(a) x^2 = 0$ for all $x \in V$. The separability of $\mathscr{S}$ then implies $P(a) = 0$. But now 1.16(i) shows that $P(g a) = 0$ for all g in the structure group G of $\mathscr{S}$. But then $(g a)^2 = 0$ for all $g \in G$, hence a is a radical element of $\mathscr{S}$, see 9.10. The semisimplicity of $\mathscr{S}$ then implies $a = 0$ which is what we wanted to prove.

10.32 Corollary. *Assume that $\mathscr{S}$ is simple and separable. Then $P(a, e) = 0$ implies $a \in K e$.*

10.33 Proposition. *Let $a \in V$, $P(a) = \operatorname{id}$.*
(i) *If $\operatorname{char}(K) \neq 2$ then there exists a central idempotent c such that $a = e + 2c$;*
(ii) *If $\operatorname{char}(K) = 2$ and if $\mathscr{S}$ is simple and separable then $a = e$.*

If $P(a) = \operatorname{id}$ then $a^2 = P(a) e = e$. If $\operatorname{char}(K) \neq 2$ put $c = \frac{1}{2}(a - e)$. Since $(e - a)^2 = e - P(a, e) a + a^2 = 2e - 2a$, we have $c^2 = c$. Moreover $P(e, c) + P(c) = 0$. With the notations of 10.3, it then follows from 10.3(ii) that $V_c' = 0$, so that c is a central idempotent.

Next let $\operatorname{char}(K) = 2$ and assume $\mathscr{S}$ to be semisimple. Let $(a_i)_{1 \leq i \leq h}$ be a maximal set of orthogonal primitive idempotents, with sum e. Let ψ be as in 10.26(iii) and let

$$V = \bigoplus_{1 \leq r \leq s \leq h} V_{rs}$$

§ 10. Peirce Decomposition Defined by an Idempotent Element

be the decomposition of 10.27. By 10.21 we have $V_{rs} = K a_r$. Let $P(a) = \mathrm{id}$ and put
$$a = \sum_{r=1}^{h} c_r a_r + \sum_{1 \leq r < s \leq h} a_{rs},$$
where $a_{rs} \in V_{rs}$. We have, using §3, (6)
$$P\left(\sum_{1}^{h} c_r t_r^2 + \sum_{1 \leq r \leq s \leq h} t_r t_s a_{rs}\right) = \psi(t_1, \ldots, t_h)^2,$$
if $t_r \in K^*$. It follows from 10.27 that
$$P(a_r, a_{st}) = 0$$
and that $c_r = 1$. But then
$$P(e, a_{st}) = \sum_{r=1}^{h} P(a_r, a_{st}) = 0.$$

If $\mathscr{S}$ is simple and separable then 10.32 implies that $a_{st} = 0$. Hence $a = e$, which proves (ii).

10.34 Corollary. *If $\mathscr{S}$ is simple and separable (if char$(K) = 2$) then $P(a) = \mathrm{id}$ implies $a = \pm e$.*

10.35 Remark. It will follow from the classification of simple J-structures, to be given in §13, that in characteristic 2 all simple J-structures are separable (see 13.7). By 10.22 the same then follows for semisimple J-structures. Hence the condition of separability in 10.31 and 10.33(ii) is redundant. It is desirable to have a more direct proof of this fact.

Notes

The Peirce decomposition is a standard tool in Jordan algebra theory, see e.g. [14, Chapter III] and [15, Chapter II]. The treatment given here is based on the interpretation of the Peirce spaces V_a, V_{e-a}, V_a' as weight spaces of the torus S_a in V (notations of 10.2 and 10.3).
10.22 is one of the basic structure theorems, due originally to Albert, in the case of Jordan algebra of characteristic not 2. The proof given here of 10.22, which makes use of properties of N, is somewhat different in spirit from the Jordan algebra proofs occuring in the literature, see e.g. [8, I, §13, p. 55–56], [14, p. 201], [15, p. 3.25].
Via the correspondence between J-structures and (quadratic) Jordan algebras, our notions of simplicity and semisimplicity of a J-structure defined over a field $k \subset K$, correspond to the notions of "absolute" simplicity and semisimplicity. Because of this, we can establish results like 10.23, which show that inseparable extensions of the base field can be avoided. It should be clear that in the framework adopted here, absolute simplicity and semisimplicity are the natural notions.

§11. Classification of Certain Algebraic Groups

This section is a preparation for the next ones, where the classification of simple J-structures is to be discussed. We shall deal with a problem in the theory of linear algebraic groups which is basic for that classification. We shall have to rely heavily on the theory of semisimple algebraic groups and their rational representations, for which we refer to [10]. For the results on root systems to be used we refer to [7].

11.1. We first show how the results of the preceding sections lead to a classification problem of algebraic groups.

Let $\mathscr{S} = (V, j, e)$ be a J-structure with structure group G. Given an idempotent element $a \neq 0, e$ in V, we can associate with it a 2-dimensional torus S_a in G (see 10.2). S_a contains the 1-dimensional subtorus C of G consisting of the scalar multiplications by elements of K^*. Let G', S' denote the identity components of the intersections of the special linear group $SL(V)$ (the subgroup of $GL(V)$ whose elements have determinant 1) with G and S_a, respectively. Clearly $G = C(G \cap SL(V))$. Moreover, G' has finite index in $G \cap SL(V)$ and CG' is connected. It follows that CG' is the identity component G° of G. Likewise we have $CS' = S_a$. The group S' is a 1-dimensional torus.

If c is a rational character of S', denote by $V(c)$ the corresponding weight space of S' in V, i.e.

$$V(c) = \{v \in V \mid s v = c(s) v \text{ for all } s \in S'\}.$$

Then V is the direct sum of the different nonzero weight spaces $V(c)$, see [1, p. 200]. It follows from the proof of 10.3 that in the present case the weight spaces are V_a, V_{e-a} and V'_a (in the notation of §10), so that at most 3 weights of S' occur. If a is not a central idempotent (and $a \neq 0, e$) there are indeed 3 weights, for all three spaces V_a, V'_a and V_{e-a} are nonzero.

Moreover, ϕ_a being the homomorphisms $(\mathbb{GL}_1)^2 \to G_1$ of 10.1, the fact that the 3 possible weights for the corresponding rational representation $(\mathbb{GL}_1)^2 \to GL(V)$ are given by $\phi_a(t, u) = t^2, tu, u^2$ (which follows

§11. Classification of Certain Algebraic Groups

from 10.3(i)) implies that the weights of S' in V have a special property. Identifying the group of rational characters of S' with $\mathbb{Z}$, there correspond to the 3 possible weights of S' three distinct integers b, c, d, which can be chosen such that $b+d=2c$. The corresponding weight spaces are V_a, V'_a, V_{e-a}, respectively. We can order the character group of S' such that $b>c>d$, $b>0$ (as is easily seen).

Finally, if a is a primitive idempotent and if $\mathscr{S}$ is semisimple, we have by 10.21 that $\dim V_a = 1$, which means that the corresponding weight b of $\mathscr{S}'$ has multiplicity 1.

We shall now study in more detail such a situation, assuming moreover that the group G' acts *irreducibly* in V. We now change the notation and we shall formulate the problem to be discussed in a precise manner.

V denotes a finite dimensional vector space. Let G be a nontrivial connected linear algebraic group, contained in the special linear group $SL(V)$. The following lemma is well-known. We sketch the proof.

11.2 Lemma. *If G acts irreducibly in V then G is semisimple.*

Let U be the unipotent radical of G. Let

$$V_0 = \{v \in V \mid U v_0 = v_0\}.$$

One knows that $V_0 \neq 0$ (see [1, p. 158]). Also, since U is a normal subgroup of G, we have that G stabilizes V_0. The irreducibility assumption now implies that $V_0 = V$, hence $U = \{e\}$.

The radical R of G is then a central torus in G. By Schur's lemma, the elements of R are scalar multiples of the identity. But since $G \subset SL(V)$, it follows that R must be finite, hence $R = \{e\}$. This establishes 11.2.

Let S be a subtorus of G. Let c be a weight of S in V, i.e. a character of S such that the corresponding weight space $V(c)$ (the definition of which was recalled in 11.1) is nonzero. Its dimension is then called the *multiplicity* of the weight c. Let $\mathfrak{g}$ be the Lie algebra of G. S acts in $\mathfrak{g}$ via the adjoint representation of G in $\mathfrak{g}$. The nonzero weights of S in $\mathfrak{g}$ are the roots of S. The following lemma is probably well-known. We consider the character group of S as an additive abelian group.

11.3 Lemma. *Any root of S is a difference of 2 weights of S in V.*

G is an algebraic subgroup of $SL(V)$, hence of $GL(V)$. Its Lie algebra $\mathfrak{g}$ is then a subalgebra of the Lie algebra $\mathfrak{gl}(V)$ of all endomorphisms of V. It is clear that it suffices to prove the assertion for $G = GL(V)$. S is contained in a maximal torus T of $GL(V)$. But the maximal tori in $GL(V)$ are the groups of nonsingular diagonal transformations, with respect to some basis of V. For such a T the assertion of the lemma is obvious.

11.4. We now assume moreover the following:
(a) G acts irreducibly in V;
(b) There exists a 1-dimensional subtorus S of G such that one of the following conditions is satisfied:
(1) S has two weights in V,
(2) S has three weights in V. There exists an isomorphism of the character group $X(S)$ of S onto $\mathbb{Z}$, such that the weights correspond to three integers a, b, c with $a > b > c$ and $a + c = 2b$ and such that the weight corresponding to a has multiplicity 1.

We call such a triple (V, G, S) *admissible*. We refer to the two possible situations in (b) as case 1 and case 2. We now can state the aim of this section. It is the solution of the following problem.

Problem. *Determine the admissible triples (V, G, S).*
Although the J-structures lead to case 2 only, the solution of the problem in case 1 is needed for the solution in case 2.

11.5. Let (V, G, S) be admissible. By 11.2 we have that G is semisimple. Let T be a maximal torus in G containing S, let $X(T)$ be its character group, let $\phi: X(T) \to X(S) = \mathbb{Z}$ be the restriction homomorphism. We choose compatible orderings on $X(T)$ and $X(S)$ (which is possible by [4, 3.1, p. 71]), such that the order on $\mathbb{Z}$ has the properties described above in case 2.

Let Σ be the root system of G with respect to T, let Δ be the basis of Σ defined by the order of $X(T)$, see [7, Ch. VI, Déf. 2, p. 153]. We say that G is *quasi-simple* if Σ is irreducible.

It is known that there exists a set $(G_i)_{1 \le i \le h}$ of quasi-simple semisimple groups G_i and a central isogeny $\alpha: G_1 \times \cdots \times G_h \to G$ (which means that α is a surjective homomorphism of algebraic groups, with finite kernel, such that the restriction of α to a maximal connected unipotent subgroup U is an isomorphism in the sense of algebraic groups of U onto $\alpha(U)$).

11.6. Let $G' = G_1 \times G_2$ be a direct product of semisimple groups. Let $\rho_i: G_i \to SL(V_i)$ be nontrivial irreducible rational representations ($i = 1, 2$). Put $V = V_1 \otimes V_2$ and let $\rho: G' \to SL(V)$ be the tensor product representation $\rho_1 \otimes \rho_2$ of G' in V. It is known that ρ is irreducible.

Now let S_1 be a 1-dimensional torus in G such that $(V_1, \rho_1(G_1), \rho_1(S))$ is admissible. Put $G = \rho(G')$, $S = \rho(S_1 \times \{e\})$. It is immediate that (V, G, S) satisfies all conditions of an admissible triple, except perhaps the multiplicity condition in case 2. At any rate (V, G, S) is admissible if we are in case 1. Such an admissible triple is called *decomposable*. Otherwise it is called *indecomposable*.

§ 11. Classification of Certain Algebraic Groups

In case 2, an admissible triple is always indecomposable. The same holds in case 1, if one of the weights of S in V has multiplicity 1. We can now state a first result on the classification of admissible triples.

11.7 Proposition. *Let (V, G, S) be admissible.*
Case 1. Suppose that (V, G, S) is indecomposable. Then G is quasi-simple.
Case 2. G is either quasi-simple or G is isogeneous to a product of two quasi-simple groups G_1 and G_2. In the second case there exist irreducible rational representations $\rho_i: G_i \to GL(V_i)$ and 1-dimensional subtori S_i in G_i such that $(V_i, \rho_i(G_i), \rho_i(S_i))$ is admissible in case 1 and that $V = V_1 \otimes V_2$, $G = \rho_1(G_1) \otimes \rho_2(G_2)$. Moreover one of the weights of S_i in V_i has multiplicity 1 ($i = 1, 2$).

Suppose that G is not quasi-simple. Let $\alpha: G_1 \times G_2 \to G$ be a central isogeny, G_1 and G_2 being semisimple groups. Let S' be the identity component of $\alpha^{-1}(S)$ and let S_i be the projection of S' on G_i. We have an irreducible rational representation ρ of $G_1 \times G_2$ in V. It is known that there exists irreducible rational representations $\rho_i: G \to GL(V_i)$ ($i = 1, 2$) such that $V = V_1 \otimes V_2$, $\rho = \rho_1 \otimes \rho_2$. Since (V, G, S) is indecomposable, we must have $S_i \neq \{1\}$.

Fix isomorphisms $\tau_i: \mathbb{GL}_1 \to S_i$ ($i = 1, 2$). Then there exists an isomorphism $\tau: \mathbb{GL}_1 \to S$ with

$$\tau(x) = \rho_1(\tau_1(x))^{\lambda_1} \otimes \rho_2(\tau_2(x))^{\lambda_2},$$

where λ_1 and λ_2 are nonzero integers.

Let v_i be a weight vector of S_i in V_i. Hence

$$\rho_i(\tau_i(x)) v_i = x^{a_i} v_i.$$

Then

$$\tau(x)(v_1 \otimes v_2) = x^{\lambda_1 a_1 + \lambda_2 a_2}(v_1 \otimes v_2),$$

so that $\lambda_1 a_1 + \lambda_2 a_2$ is a weight of $\mathbb{GL}_1$ in V. If v_i' is a second weight vector of S_i in V_i, with a different weight, we would get weights $\lambda_1 a_1 + \lambda_2 a_2$, $\lambda_1 a_1' + \lambda_2 a_2$, $\lambda_1 a_1 + \lambda_2 a_2'$, $\lambda_1 a_1' + \lambda_2 a_2'$ of $\mathbb{GL}_1$ in V, with $a_1 \neq a_1', a_2 \neq a_2'$. It is immediate that among these integers there are at least 3 distinct ones. It follows that, if S is to have 2 weights in V, at least one S_i has only one weight in V_i hence would be in the centre of G_i, which is absurd. Hence the first assertion follows.

A similar argument shows that in case 2, S_i can have only 2 weights in V_i. The multiplicity condition in case 2 implies that at least one weight of S_i must have multiplicity 1 in V_i. This establishes the assertion about case 2.

11.8. Let (V, G, S) be admissible, with G quasi-simple. We use the notations of 11.5.

Let $\tilde{\alpha}$ be the highest root of Σ, with respect to the ordering which we have chosen (see [7, p. 166]).
We have
$$\tilde{\alpha} = \sum_{\alpha \in \Delta} h_\alpha \alpha.$$

The positive integer h_α is called the coefficient of α in $\tilde{\alpha}$, we have $h_\alpha \geq 1$. Let ϕ be as in 11.5.

11.9 Lemma. *There exists an integer $n \geq 1$ such that $\phi(\Sigma) \subset \{n, 0, -n\}$ in case 1, respectively $\phi(\Sigma) \subset \{2n, n, 0, -n, -2n\}$ in case 2.*

Let a, b (respectively a, b, c) be the weights of S in V. In case 2 assume them to be as in 11.4. The assertion now follows from 11.3, with $n = |a-b|$.

11.10 Lemma. (i) *In case 1 there exists exactly on $\alpha \in \Delta$ with $\phi(\alpha) \neq 0$. It has coefficient 1 in $\tilde{\alpha}$;*
(ii) *In case 2 there are at most two $\alpha \in \Delta$ with $\phi(\alpha) \neq 0$. They have coefficient ≤ 2 in $\tilde{\alpha}$.*

We have $\phi(\tilde{\alpha}) = \sum_{\alpha \in \Delta} h_\alpha \phi(\alpha)$. n being as in 11.9 we have in case 1 that $\phi(\tilde{\alpha}) = n$, $\phi(\alpha) = 0$ or n. (i) follows at once from this observation. The proof of (ii) is similar.

Remark. It follows from 11.10 that in case 1 only such quasi-simple G can occur, for which a coefficient 1 occurs in the highest root. This already rules out groups of type E_8, F_4, G_2, where all coefficients h_α are ≥ 2, see [7, pp. 269, 272, 274].

11.11. Let W be the Weyl group of T. W acts in the real vector space $R = X(T) \otimes_\mathbb{Z} \mathbb{R}$. Let $(\,|\,)$ be a positive definite scalar product on $R \times R$ which is W-invariant. We identify Σ with a subset of R.

The fundamental weights ϖ_α ($\alpha \in \Delta$) are defined by
$$2(\varpi_\alpha | \beta)(\beta | \beta)^{-1} = \delta_{\alpha\beta} \qquad (\alpha, \beta \in \Delta)$$

where $\delta_{\alpha\beta}$ is the Kronecker symbol. Let P be the subgroup of R generated by the ϖ_α ($\alpha \in \Delta$). We order the elements of P as follows: $x < y$ if $y - x$ is a sum of positive roots.
There exist a unique highest weight (for this order) $\varpi \in P \cap X(T)$ of T in V, with multiplicity 1. We have

(1) $$\varpi = \sum_{\alpha \in \Delta} n_\alpha \varpi_\alpha$$

where the n_α are integers ≥ 0 (see [10, exposé 16]).

§11. Classification of Certain Algebraic Groups

Let $w_0 \in W$ be the element which sends positive roots into negative ones. We put $x^* = w_0 x$ for all $x \in P$. Then $w_0 \varpi = \varpi^*$ is the lowest weight of T in V, as follows from the definitions.

11.12 Lemma. *Assume that Σ is irreducible, let $\alpha \in \Delta$.*
(i) *There exist strictly positive rational numbers $n_{\alpha\beta}$ such that*

(2) $$\varpi_\alpha = \sum_{\beta \in \Delta} n_{\alpha\beta} \beta;$$

(ii) *There exist strictly positive integers $m_{\alpha\beta}$ such that*

$$\varpi_\alpha - \varpi_\alpha^* = \sum_{\beta \in \Delta} m_{\alpha\beta} \beta.$$

This can be proved by checking cases, using the tables of [7, pp. 250–275]. A better proof is as follows. It is clear that there exist rational numbers $n_{\alpha\beta}$ such that (2) holds. That $n_{\alpha\beta} \geq 0$ follows from the following facts: (a) $(\beta, \gamma) \leq 0$ if $\beta, \gamma \in \Delta$, $\beta \neq \gamma$, (b) $(\varpi_\alpha, \varpi_\beta) \geq 0$ for all $\beta \in \Delta$, see [7, Ch. V, §3, Lemme 6, p. 79].

Now assume that $n_{\alpha\gamma} = 0$ for some $\gamma \in \Delta$. Let s_γ be the reflection defined by γ, i.e. the linear transformations of R given by

$$s_\gamma x = x - 2(\gamma|\gamma)^{-1}(\gamma|x)\gamma.$$

Then

$$s_\gamma \varpi_\alpha = \varpi_\alpha + n\gamma$$

where

$$n = -\sum_{\beta \neq \gamma} 2 n_{\alpha\beta} (\gamma|\gamma)^{-1} (\gamma|\beta).$$

From $(s_\gamma \varpi_\alpha | s_\gamma \varpi_\alpha) = (\varpi_\alpha | \varpi_\alpha)$ it follows that we must have $n = 0$. Since $n_{\alpha\beta} \geq 0$, $(\gamma|\beta) \leq 0$ if $\beta \neq \gamma$, it follows that $n_{\alpha\beta} = 0$ if $(\beta|\gamma) \neq 0$. The connectedness of the Dynkin graph of Σ then implies that we have $n_{\alpha\beta} = 0$ for all $\beta \in \Delta$, which is impossible. This establishes (i).

To prove (ii) write w_0 as a product of s_β $(\beta \in \Delta)$. One then sees that $\varpi_\alpha - \varpi_\alpha^*$ is an integral linear combination of the $\beta \in \Delta$. (ii) then follows from (i), observing that there is a permutation π of Δ such that $w_0 \beta = -\pi(\beta)$.

11.13 Lemma. *There exist strictly positive integers m_β $(\beta \in \Delta)$ such that*

$$\varpi - \varpi^* = \sum_{\beta \in \Delta} m_\beta \beta.$$

If $\varpi \neq \varpi_\alpha$ for all $\alpha \in \Delta$ then $m_\beta \geq 2$ for all $\beta \in \Delta$.

This follows from 11.12, using (1).

We can now establish a result which will allows us to give the full classification of admissible (V, G, S). Recall that by 11.10 there are at most 2 roots $\alpha \in \Delta$ with $\phi(\alpha) \neq 0$. If we are in case 1 there is only one.

11.14 Proposition. *Let (V, G, S) be admissible with G quasi-simple.*
(A) *Assume that there is one $\alpha \in \Delta$ with $\phi(\alpha) \neq 0$.*
(i) *S is the identity component S_α of the intersection of the kernels of the $\beta \in \Delta$ with $\beta \neq \alpha$;*
(ii) *In case 1 there exists $\beta \in \Delta$ such that $\varpi = \varpi_\beta$. α has coefficient 1 in $\varpi_\beta - \varpi_\beta^*$ and in $\tilde{\alpha}$. If one of the weights of S in V has multiplicity 1 then $\beta = \alpha$;*
(iii) *In case 2 we have $\varpi = \varpi_\alpha$ or $\varpi = 2\varpi_\alpha$. If $\varpi = \varpi_\alpha$ then α has coefficient 2 in $\varpi_\alpha - \varpi_\alpha^*$. If $\varpi = 2\varpi_\alpha$ then α has coefficient ≤ 2 in $\tilde{\alpha}$.*
(B) *Assume that there are two roots $\alpha, \beta \in \Delta$ with $\phi(\alpha) \neq 0$, $\phi(\beta) \neq 0$.*
(iv) *We then have $\varpi = \varpi_\alpha$ or $\varpi = \varpi_\beta$, moreover α and β have coefficient 1 in $\varpi - \varpi^*$.*

First consider case 1, so that we have the situation (A). Let a and b ($a > b$) be the weights of S in V ($X(S)$ being identified with $\mathbb{Z}$).

Let $\alpha \in \Delta$, $\phi(\alpha) \neq 0$. By 11.3 we have $\phi(\alpha) = a - b$. But clearly $\phi(\varpi) = a$, $\phi(\varpi^*) = b$, hence also $\phi(\varpi - \varpi^*) = a - b = \phi(\alpha)$. By 11.13 it follows that

$$(3) \qquad \phi(\varpi - \varpi^*) = \sum_{\beta \in \Delta} m_\beta \, \phi(\beta),$$

whence $a - b = m_\alpha(a - b)$. Using 11.13 the first assertion of (ii) follows. It also follows that α must have coefficient 1 in $\varpi_\beta - \varpi_\beta^*$. That α has coefficient 1 in $\tilde{\alpha}$ was already proved in 11.10(i).

We next prove (i). Clearly S is contained in S_α. Since both S and S_α are 1-dimensional tori they must be equal.

In case 1 it remains to prove the last point of (ii). Let $s_\gamma \in W$ be the reflection defined by $\gamma \in \Delta$. Then

$$s_\gamma \varpi = \varpi - 2(\varpi|\gamma)(\gamma|\gamma)^{-1} \gamma$$

is again a weight of T in V and

$$\phi(s_\gamma \varpi) = \phi(\varpi) - 2(\varpi|\gamma)(\gamma|\gamma)^{-1} \phi(\gamma).$$

If one of the weights of S in V, say the highest one a, has multiplicity 1, it follows that $(\varpi|\gamma) \neq 0$ can only be if $\phi(\gamma) \neq 0$. This shows that we must have $\beta = \alpha$ in (ii), as asserted.

New assume that we are in case 2 and in the situation (A). Let a, b, c be the weights of S in V, with $a > b > c$, $2b = a - c$. Then $\phi(\varpi - \varpi^*) = a - c = 2(a - b)$. By 11.3 we have $\phi(\alpha) \geq a - b$ or $\phi(\alpha) = 0$, if $\alpha \in \Delta$. 11.13 now implies that

$$\sum_{\alpha \in \Delta, \phi(\alpha) \neq 0} m_\alpha \leq 2.$$

The multiplicity condition shows, as before, that $(\varpi|\alpha) \neq 0$ implies $\phi(\alpha) \neq 0$. Let $\phi(\alpha) \neq 0$, $m_\alpha = 1$. By 11.13 there is $\beta \in \Delta$ with $\varpi = \varpi_\beta$, moreover $(\varpi_\beta|\alpha) \neq 0$, wence $\beta = \alpha$.

§ 11. Classification of Certain Algebraic Groups

If $\phi(\alpha) \neq 0$, $m_\alpha = 2$ the inequality shows that $\phi(\beta) = 0$ for $\beta \neq \alpha$ whence $\varpi = n\varpi_\alpha$. 11.13 implies that then we must have $n \leq 2$. (i) has now been established.

It has been proved that $\varpi = \varpi_\alpha$ or $\varpi = 2\varpi_\alpha$ and that α has coefficient ≤ 2 in $\varpi_\alpha - \varpi_\alpha^*$. Now assume that $\varpi = \varpi_\alpha$ and that α has coefficient 1 in $\varpi_\alpha - \varpi_\alpha^*$. If c is a weight of T in V then we have

$$\varpi_\alpha - c = \sum_{\beta \in \Delta} k_\beta \beta,$$

where the k_β are positive integers. Since ϖ_α^* is the lowest weight our assumption implies that $k_\alpha \leq 1$. Hence

$$\phi(c) = \phi(\varpi_\alpha) - k_\alpha \phi(\alpha),$$

and it would follow that S had only 2 weights in V, which is a contradiction. Hence α has coefficient 2 in $\varpi_\alpha - \varpi_\alpha^*$. It is also clear that, if $\varpi = 2\varpi_\alpha$, α must have coefficient 1 in $\varpi_\alpha - \varpi_\alpha^*$. The last assertion of (iii) was already established in 11.10(ii). We finally have to deal with (B). To prove (iv) one uses similar arguments. We leave the details to the reader.

A corollary of the proof is a converse of 11.14 in case 1. We now denote by G a quasi-simple linear group, which we assume to be simply connected (for this notion see [2, p. 193-194]). The notations remain the same. It is known that for any ϖ of the form (1) there exists an irreducible rational representations ρ of G such that T has highest weight ϖ in that representation. If $\alpha \in \Delta$, we denote again by S_α the identity component of the intersection of the kernels of all $\beta \in \Delta$ with $\beta \neq \alpha$.

11.15 Corollary. *Let $\alpha, \beta \in \Delta$. Suppose that α has coefficient 1 in $\varpi_\beta - \varpi_\beta^*$. Let $\rho: G \to GL(V)$ be an irreducible rational representation such that T has highest weight ϖ_β in V. Then $(V, \rho(G), \rho(S_\alpha))$ is admissible in case 1.*

To prove that $\rho(S_\alpha)$ has 2 weights in V one proceeds in the same way as in the last part of the proof of 11.14.

In case 2 a similar result is true (see 11.18). But we prefer to establish this via the explicit classification of admissible triples, to which we turn now.

11.16 Classification. 11.14 reduces the determination of the admissible (V, G, S) with G quasi-simple to the solution of a problem on root systems, viz. the determination in the situation of 11.14(A) of the possible $\alpha \in \Delta$ and ϖ_β such that the conditions of 11.14(i) (respectively, 11.14(ii)) are fulfilled and the determination when the situation of 11.14(iv) can hold.

The root system Σ of G with respect to T is of one of the types

$$A_l(l \geq 1), \ B_l(l \geq 2), \ C_l(l \geq 3), \ D_l(l \geq 4), \ E_6, \ E_7, \ E_8, \ F_4, \ G_2.$$

We shall check the possibilities for the various types.

We use the notations of [7, p. 250–275] for the explicit description of the irreducible root systems and the numbering of simple roots and fundamental weights. If α_i is a simple root we write ϖ_i instead of ϖ_{α_i}. As in 11.11, w_0 denotes the element of the Weyl group W which sends positive roots into negative ones.

Type A_l. We have

$$\varpi_i = (l+1)^{-1}[(l-i+1)\alpha_1 + 2(l-i+1)\alpha_2 + \cdots + (i-1)(l-i+1)\alpha_{i-1}$$
$$+ i(l-i+1)\alpha_i + i(l-i)\alpha_{i+1} + \cdots + i\alpha_l].$$

Since $w_0 \alpha_i = -\alpha_{l-i+1}$, we find that

$$\varpi_i - \varpi_i^* = \alpha_1 + 2\alpha_2 + \cdots + (i-1)\alpha_{i-1} + i(\alpha_i + \cdots + \alpha_{l-i+1})$$
$$+ (i-1)\alpha_{l-i+2} + \cdots + 2\alpha_{l-1} + \alpha_l,$$

if $i \leq l-i+1$, moreover

$$\varpi_i - \varpi_i^* = \varpi_{l-i+1} - \varpi_{l-i+1}^* \quad (1 \leq i \leq l).$$

In case 1, we must have i and j such that α_i has coefficient 1 in $\varpi_j - \varpi_j^*$. The preceding formulas show that we must have one of the following cases for α and ϖ

$$\alpha = \alpha_1 \text{ or } \alpha_l, \quad \varpi = \varpi_j \quad (1 < j < l);$$
$$\alpha = \alpha_i \quad (1 \leq i \leq l), \quad \varpi = \varpi_1 \text{ or } \varpi_l.$$

Assume now that we are in case 2 and that there is only one $\alpha \in \Delta$ with $\phi(\alpha) \neq 0$. We then must have i such that α_i has coefficient 1 or 2 in ϖ_i. We obtain the following possibilities:

$$\alpha = \alpha_1, \quad \varpi = 2\varpi_1; \quad \alpha = \alpha_l, \quad \varpi = 2\varpi_l;$$
$$\alpha = \alpha_2, \quad \varpi = \varpi_2; \quad \alpha = \alpha_{l-1}, \quad \varpi = \varpi_{l-1}.$$

If we are in case 2 and if there are two roots $\alpha, \beta \in \Delta$ with $\phi(\alpha) \neq 0$, $\phi(\beta) \neq 0$, then 11.14(iv) implies that we can only have

$$\alpha = \alpha_1, \quad \beta = \alpha_l, \quad \varpi = \varpi_1 \text{ or } \varpi_l.$$

We next discuss the other simple types. It will follows from 11.14(iv) and the formulas for the $\varpi_i - \varpi_i^*$, given below, that there no two distinct $\alpha, \beta \in \Delta$ can have coefficient 1 in $\varpi_i - \varpi_i^*$, so that the situation (B) of 11.14 is impossible. We therefore assume from now on that we have the situation of 11.14(A).

§11. Classification of Certain Algebraic Groups   115

Type B_l. We have $w_0 = -1$, $\varpi_i^* = -\varpi_i$. Then
$$\varpi_i - \varpi_i^* = 2\alpha_1 + 4\alpha_2 + \cdots + 2(i-1)\alpha_{i-1} + 2i(\alpha_i + \cdots + \alpha_l) \quad (1 \leq i \leq l),$$
$$\varpi_l - \varpi_l^* = \alpha_1 + 2\alpha_2 + \cdots + l\alpha_l.$$

In case 1 the only possibility is $\alpha = \alpha_1$, $\varpi = \varpi_l$. In case 2 we must have $\alpha = \alpha_1$, $\varpi = \varpi_1$ or $l = 2$, $\alpha = \alpha_2$, $\varpi = \varpi_2$.

Type C_l. $w_0 = -1$, $\varpi_i^* = -\varpi_i$ and
$$\varpi_i - \varpi_i^* = 2\alpha_1 + 4\alpha_2 + \cdots + 2(i-1)\alpha_{i-1} + i(2\alpha_i + 2\alpha_{i+1} + \cdots + 2\alpha_{l-1} + \alpha_l).$$

In case 1 the only possibility is $\alpha = \alpha_l$, $\varpi = \varpi_1$. In case 2 we must have $\alpha = \alpha_1$, $\varpi = \varpi_l$.

Type D_l. We have $w_0 = -1$ if l is even. If l is odd then
$$w_0 \alpha_i = -\alpha_i \quad (1 \leq i \leq l-2), \quad w_0(\alpha_{l-1}) = -\alpha_l, \quad w_0(\alpha_l) = -\alpha_{l-1}.$$

We find from the formulas of [loc. cit., p. 256–257] that
$$\varpi_i - \varpi_i^* = 2\alpha_1 + 4\alpha_2 + \cdots + 2(i-1)\alpha_{i-1}$$
$$+ 2i(\alpha_i + \cdots + \alpha_{l-2}) + i(\alpha_{l-1} + \alpha_l) \quad (1 \leq i \leq l-2),$$
$$\varpi_{l-1} - \varpi_{l-1}^* = \alpha_1 + 2\alpha_2 + \cdots + (l-2)\alpha_{l-2}$$
$$+ \tfrac{1}{2} l \alpha_{l-1} + \tfrac{1}{2}(l-2)\alpha_l \quad \text{if } l \text{ is even,}$$
$$\varpi_{l-1} - \varpi_{l-1}^* = \alpha_1 + 2\alpha_2 + \cdots + (l-2)\alpha_{l-2}$$
$$+ \tfrac{1}{2}(l-1)(\alpha_{l-1} + \alpha_l) \quad \text{if } l \text{ is odd,}$$

and $\varpi_l - \varpi_l^*$ is obtained from $\varpi_{l-1} - \varpi_{l-1}^*$ by permuting α_l and α_{l-1}. It follows that in case 1 we have the following possibilities for $l > 4$
$$\alpha = \alpha_1, \quad \varpi = \varpi_l \text{ or } \varpi_{l-1};$$
$$\alpha = \alpha_l \text{ or } \alpha_{l-1}, \quad \varpi = \varpi_1.$$

If $l = 4$ we can also have
$$\alpha = \alpha_3, \quad \varpi = \varpi_4; \quad \alpha = \alpha_4, \quad \varpi = \varpi_3.$$

In case 2 the only possibility is $\alpha = \alpha_1$, $\varpi = \varpi_1$ if $l > 5$. If $l = 4$ we can also have
$$\alpha = \alpha_3, \quad \varpi = \varpi_3; \quad \alpha = \alpha_4, \quad \varpi = \varpi_4.$$

If $l = 5$ there is also the possibility that $\alpha = \alpha_4$ or α_5, $\varpi = \varpi_4$ or ϖ_5.

Type E_6. Making use of the results of [7, p. 261] one sees that the coefficient of α_i in $\varpi_j - \varpi_j^*$ is always ≥ 2 and that a coefficient 2 only arises if $i = 1, 6$ and $j = 1, 2, 6$. This shows that case 1 is impossible and that in case 2 we can only have $\alpha = \alpha_1$, $\varpi = \varpi_1$ or $\alpha = \alpha_6$, $\varpi = \varpi_6$.

Type E_7. We have $w_0 = -1$. All coefficients of the α_i in ϖ_j are >1, except the coefficient of α_7 in ϖ_1, as follows from [loc. cit., p. 265]. Hence both case 1 and case 2 are impossible.

The remark after 11.10 shows that in the remaining types E_8, F_4, G_2 case 1 cannot occur. Next consider case 2. We only deal with type E_8, in the other types F_4 and G_2 the argument is the same. By 11.14(iii), α has to have coefficient ≤ 2 in the highest root $\tilde{\alpha}$. By [loc. cit., p. 269] we must have $\alpha = \alpha_1$ or $\alpha = \alpha_8$. But since $w_0 = -1$ and since α_i has coefficient ≥ 2 in ϖ_i ($i = 1, 8$) it follows that case 2 is impossible.

It also follows from the preceeding discussion that if (V, G, S) is admissible in case 1 and if S has a weight with multiplicity 1 in V, then (by 11.14(ii), last point) we must have type A_l, $\alpha = \alpha_1$, $\varpi = \varpi_1$ or $\alpha = \alpha_l$, $\varpi = \varpi_l$.

11.17 Explicit description of admissible (V, G, S) **(G quasi-simple).** Except for a possibility with type E_6, which occurs in case 2, all possibilities listed above provide "classical groups", i.e. of types A_l, B_l, C_l, D_l. Then explicit descriptions of the groups and their representations exist, which lead to concrete realizations of the various possibilities.

We denote by $\tilde{G}$ a simply connected quasi-simple group of the same type as G. Let $\pi: \tilde{G} \to G$ be a central isogeny. Let $\tilde{T}$ be a maximal torus in $\tilde{G}$, with $\pi(\tilde{T}) = T$. The root system of $\tilde{G}$ with respect to $\tilde{T}$ can be identified with Σ.

If ϖ_i is a fundamental weight, let ρ_i denote an irreducible rational representation of $\tilde{G}$ such that T has highest weight ϖ_i in ρ_i. Such a ρ_i exists.

Let S_i be the identity component of the intersection of the kernels in $\tilde{T}$ of the α_j ($j \neq i$). The discussion of 11.16, together with 11.14 now leads to the following explicit descriptions of the admissible triples.

Type A_l. We may take $\tilde{G} = \mathbb{SL}_{l+1}$. Let $A = K^{l+1}$, let $(e_i)_{1 \leq i \leq l+1}$ be the canonical basis of A. We may take $\tilde{T}$ to be the maximal torus in $\tilde{G}$ consisting of the linear transformations t of A with

$$t e_i = x_i e_i \quad (1 \leq i \leq l+1),$$

with $\prod_{i=1}^{l+1} x_i = 1$. Define a character ε_i of T by

$$\varepsilon_i(t) = x_i.$$

Then the ε_i are as in [7, p. 250-251], as follows e.g. from [10, exp. 20].

It is known (see [10, p. 20-08]) that ρ_i is the canonical representation of G in the exterior power $V_i = \overset{i}{\wedge} A$ ($1 \leq i \leq l$). Let S_i be the 1-dimensional

§ 11. Classification of Certain Algebraic Groups

torus consisting of the $t \in T$ with

$$t e_j = x^{l+1-i} e_j \quad (1 \leq j \leq i),$$
$$t e_j = x^{-i} e_j \quad (i+1 \leq j \leq l+1),$$

with $x \in K^*$.

The admissible triples in case 1 are

$$(V_i, \rho_i(\mathbb{SL}_{l+1}), \rho_i(S_j)),$$

with either $1 < i < l, j = i, l$ or $1 \leq j \leq l, i = 1, l$.
If $\rho_i(S_j)$ has to have a weight with multiplicity 1, the only possibilities are $i = j = 1$ or l (see the end of 11.16). Since $\rho_1 = \mathrm{id}$ and since ρ_l is the contragredient of ρ_1, there is then essentially only one possibility, viz. $(K^{l+1}, \mathbb{SL}_{l+1}, S_1)$.

Next consider case 2. If $\mathrm{char}(K) \neq 2$ the representation σ with highest weight $2\varpi_1$ is equivalent to the representation of $\mathbb{SL}_{l+1}$ in the space $\mathbb{S}_{l+1}$ of all symmetric $(l+1) \times (l+1)$ matrices, given by $\sigma(X) S = XS \cdot {}^t X (X \in \mathbb{SL}_{l+1}(K), S \in \mathbb{S}_{l+1})$.
It is easily checked that the torus S_1 has in that case in fact 3 weights in $\mathbb{S}_{l+1}$, with multiplicities $1, l, \frac{1}{2} l(l+1)$, respectively. This gives the admissible triple

$$(\mathbb{S}_{l+1}, \sigma(\mathbb{SL}_{l+1}), \sigma(S_1)).$$

The representation with highest weight $2\varpi_l$ leads to the same admissible triple.

If $\mathrm{char}(K) = 2$, the irreducible representation with highest weight $2\varpi_1$ is the transform under the Frobenius endomorphism of the one with highest weight ϖ_1 (see [2, first paragraph of p. 52]), which implies that then $\sigma(S_1)$ has only two weights in the space of its representation. So this case is impossible, likewise that of highest weight $2\varpi_l$.

The next possibilities for case 2 in type A_l lead to the irreducible representation α of $\mathbb{SL}_{l+1}$ in the space $\mathbb{A}_{l+1}$ of all alternating $(l+1) \times (l+1)$ matrices. One checks easily that S_2 has 3 weights in that representation with multiplicaties $1, 2(l-1), \frac{1}{2}(l-1)(l-2)$, respectively.
This gives the admissible triple

(4) $$(\mathbb{A}_{l+1}, \alpha(\mathbb{SL}_{l+1}), \alpha(S_2)).$$

The final case in type A_l, where there are 2 roots $\alpha, \beta \in \Delta$ with $\phi(\alpha) \neq 0, \phi(\beta) \neq 0$ gives an admissible triple

(5) $$(K^l, \mathbb{SL}_l, S),$$

where S is a suitable 1-dimensional torus (which is not unique).

From the discussion in 11.16 it follows that in the remaining possibilities for case 1 we find the symplectic groups in their standard representations, the even dimensional orthogonal groups in their ordinary representations and the various spin groups, in their spin representations. We do not insist on the details, which will not be needed in the sequel.

From now on we discuss the remaining possibilities for case 2. If $\text{char}(K) \neq 2$, we denote by $\mathfrak{SO}_n$ the special orthogonal group, i.e. the subgroup of $\mathfrak{SL}_n$ consisting of the matrices X with $X \cdot {}^t X = 1$. If $\text{char}(K) = 2$ and n is even, then an analogue of $\mathfrak{SO}_n$ can also be defined (see [12, p. 65]). The simply connected covering group $\tilde{G}$ of $G = \mathfrak{SO}_n$ is the spin group $\mathfrak{Spin}_n$. The symplectic group $\mathfrak{Sp}_{2n}$ is the subgroup of $\mathfrak{SL}_{2n}$ consisting of the X with ${}^t X \begin{pmatrix} 0 & 1_n \\ -1_n & 0 \end{pmatrix} X = 1_{2n}$ (where 1_r denotes an identity matrix). $\mathfrak{Sp}_{2n}$ is simply connected.

Type B_l. First let $\text{char}(K) \neq 2$. We then have $\tilde{G} = \mathfrak{Spin}_{2l+1}$. The irreducible representation occurring in case 2 is ρ_1, with highest weight ϖ_1. We can identify $\rho_1(G)$ with $\mathfrak{SO}_{2l+1}$.

Let $V = K^{2l+1}$, let $(e_i)_{0 \leq i \leq 2l}$ be its canonical basis. Define a quadratic form Q on V by

$$Q\left(\sum_{i=0}^{2l} x_i e_i\right) = x_0^2 + \sum_{i=1}^{l} x_i x_{i+l}.$$

We may take for T the torus in G consisting of the linear transformations t with

$$t e_0 = e_0,$$
$$t e_i = x_i e_i,$$
$$t e_{i+l} = x_i^{-1} e_{i+l},$$

where $1 \leq i \leq l$, $x_i \in K^*$. Let ε_i be the character of $\tilde{T}$ with

$$\varepsilon_i(t) = x_i \quad (1 \leq i \leq l).$$

The ε_i are as in [7, p. 252-253], see [10, p. 22-04].

Let S be the 1-dimensional torus with $s e_i = e_i$ ($i \neq 1, l+1$), $s e_1 = x e_1$, $s e_{l+1} = x^{-1} e_{l+1}$ ($x \in K^*$). It is immediate that S has only 3 weights in V, with multiplicities $1, 1, 2l-1$, respectively. We obtain the admissible triple $(K^{2l+1}, \mathfrak{SO}_{2l+1}, S)$.

If $\text{char}(K) = 2$, we may take $\tilde{G}$ to be the symplectic group $\mathfrak{Sp}_{2l}$ (see [10, p. 22-04, 22-05]). We shall recover this case in type C_l. Also, the possibility $l=2$, $\alpha = \alpha_2$, $\varpi = \varpi_2$ for case 2 can be recovered in type C_2 (via the isomorphism of groups of type B_2 and groups of type C_2).

Type D_l. $\tilde{G}$ is now the spin group $\mathfrak{Spin}_{2l}$. The irreducible representation which we deal with first is ρ_1, with highest weight ϖ_1. We then may take $G = \mathfrak{SO}_{2l}$. Let $V = K^{2l}$, let $(e_i)_{1 \leq i \leq l}$ be its canonical basis. Let Q be the

§ 11. Classification of Certain Algebraic Groups

quadratic form on V with

$$Q\left(\sum_{i=1}^{2l} x_i e_i\right) = \sum_{i=1}^{l} x_i x_{i+l}.$$

Let S be the torus consisting of the linear transformations s with $se_1 = xe_1$, $se_{l+1} = x^{-1} e_{l+1}$, $se_i = e_i$ ($i \neq 1, l+1$), where $x \in K^*$. It has 3 weights in V, with multiplicities $1, 1, 2l-2$, respectively. The admissible triple is $(K^{2l}, \mathbb{SO}_{2l}, S)$.

The other possibilities of 11.16 in case D_4 are related to this one via the triality automorphism and do not lead to an essentially new admissible triple.

The extra possibility in type D_5 which was found in 11.16 is ruled out by observing that in that case the torus S_α of 11.14 has more than 3 weights in the representation ϖ_α (which readily follows by using the formulas of [7, p. 256–257] and the results of [10, p. 20-05]).

Type C_l ($l \geq 2$). We have $\tilde{G} = \mathbb{Sp}_{2l}$. The representation with highest weight ϖ_1 is the standard one. Let again $V = K^{2l}$, let $(e_i)_{1 \leq i \leq 2l}$ be the canonical basis. We introduce a nondegenerate alternating bilinear form $\langle\,,\,\rangle$ on $V \times V$ by

$$\left\langle \sum_{i=1}^{2l} x_i e_i, \sum_{i=1}^{2l} y_i e_i \right\rangle = \sum_{i=1}^{l} (x_i y_{i+l} - x_{i+l} y_i).$$

$\tilde{G}$ is the group of linear transformations of V leaving this form invariant.

S being as in case D_l, we obtain the admissible triple

(6) $\qquad\qquad\qquad (K^{2l}, \mathbb{Sp}_{2l}, S).$

In the remaining possibilities for case 2, where G is of type E_6, it is known that the irreducible representations which are involved, those with highest weights ϖ_1 and ϖ_6, are of dimension 27, see [10, p. 20-05]. Moreover one is the contragredient of the other. This follows from [loc.cit., p. 20-10], using what w_0 is in this case, see [7, p. 261, last 2 lines]. Let ρ be the representation with highest weight ϖ_1. It is known that $\rho(\tilde{G})$ is isomorphic to G, so that we may take $G = \tilde{G}$, the simply connected group of type E_6.

An explicit check shows that the 1-dimensional torus (which is given by 11.14(i)) has 3 distinct weights, with multiplicities 1, 10, 16, respectively (the weights of ϖ_1 are explicitly given in [10, p. 20-05]). We call the resulting admissible triples the *exceptional* ones. Since the two representations involved are contragredient, it follows that there is essentially only one such admissible triple.

We have now, via a case by case check, also proved the following partial converse of 11.14 for case 2. The notations are as in 11.14.

11.18 Proposition. *Let $\alpha \in \Delta$. Suppose either that α has coefficient 2 in $\varpi_\alpha - \varpi_\alpha^*$ or that $\mathrm{char}(K) \neq 2$ and α has coefficient 1 in $\varpi_\alpha - \varpi_\alpha^*$. Let $\rho: G \to GL(V)$ be an irreducible rational representation such that T has highest weight ϖ_α (respectively $2\varpi_\alpha$) in V. Then $(V, \rho(G), \rho(S_\alpha))$ is admissible and in case 2.*

11.19. Let (V, G, S) be admissible with G not quasi-simple. This case still remains to be dealt with. We use the notations of 11.17. The results of 11.7 together with the remark made in 11.17, at the end of the discussion of type A_l (case 1), shows that we must have in that case that $G_1 \simeq \mathbb{SL}_m$, $G_2 \simeq \mathbb{SL}_n$, $V_1 = K^m$, $V_2 = K^n$. Hence $V = K^{mn}$. Instead we take $V = \mathbb{M}_{m,n}$, the space of $m \times n$ matrices. It then readily follows that we may take now $G = \mathbb{SL}_m \times \mathbb{SL}_n$, acting in $\mathbb{M}_{m,n}$ via the representation ρ with

$$\rho(X, Y) Z = XYZ^{-1},$$

where $X \in \mathbb{SL}_m$, $Y \in \mathbb{SL}_n$, $Z \in \mathbb{M}_{m,n}$.

Hence in the present situation we have as possible admissible triples

(7) $\qquad \left(\mathbb{M}_{m,n}, \rho(\mathbb{SL}_m \times \mathbb{SL}_n), \rho(S)\right),$

where S is a certain 1-dimensional subtorus of $\mathbb{SL}_m \times \mathbb{SL}_n$.

Let (e_{ij}) $(1 \leq i \leq m, 1 \leq j \leq m)$ be the canonical basis of $\mathbb{M}_{m,n}$. Then it is readily seen that one can take $\rho(S)$ to be the torus consisting of the linear transformations s of $\mathbb{M}_{m,n}$ with

$$se_{11} = x^{2mn-m-n} e_{11},$$
$$se_{ij} = x^{mn-m-n} e_{ij} \quad \text{if } i=1, \text{ or } j=1 \text{ and } (i,j) \neq (1,1),$$
$$se_{ij} = x^{-m-n} e_{ij} \quad \text{otherwise.}$$

We finally establish the result which will be the basis of the classification of simple J-structures. Let (V, G, S) be admissible in case 2. Let C be the 1-dimensional torus of nonzero scalar multiplications in V.

11.20 Proposition. *Assume that GC has an open orbit in V, whose complement contains a hypersurface. We then have one of the following possibilities for G and V (up to isomorphism):*
(a) $V = \mathbb{M}_r$, the space of $r \times r$ matrices $(r > 1)$. G is the group of linear transformations $Z \mapsto XZY^{-1}$ of $\mathbb{M}_r$, with $X, Y \in \mathbb{SL}_r$;
(b) $\mathrm{char}\, K \neq 2$, $V = \mathbb{S}_r$, the space of symmetric $r \times r$ matrices $(r > 1)$. G is the group of linear transformations $Y \mapsto XY({}^tX)$ of $\mathbb{S}_r$, with $X \in \mathbb{SL}_r$;
(c) $V = \mathbb{A}_{2r}$, the space of alternating $2r \times 2r$ matrices $(r \geq 2)$. G is the group of linear transformations $Y \to XY({}^tX)$ of $\mathbb{A}_{2r}$, with $X \in \mathbb{SL}_{2r}$;
(d) $V = K^{2r}$, $G = \mathbb{SO}_{2r} (r \geq 3)$;

(e) char $K \neq 2$, $V = K^{2r+1}$, $G = \mathbb{SO}_{2r+1}$ ($r \geq 2$);
(f) dim $V = 27$ and G is a simply connected quasi-simple group of type E_6.

We have to prove that the condition on the orbit of GC rules out the following kinds of admissible triples: those of the form (5) and (6) and those of the form (4) with $l+1$ odd or of the form (7) with $m \neq n$. From the fact that $\mathbb{SL}_l$ acts transitively on all nonzero vectors of K^l it follows that the orbit condition does not hold for the admissible triple (5). A similar argument rules out (6).

Next consider the last case. We may assume that $m > n$. There is an open orbit U of $\mathbb{SL}_m \times \mathbb{SL}_n$ in $\mathbb{M}_{m,n}$, whose elements are the matrices with maximal rank n. The complement F of U is a closed set in $\mathbb{M}_{m,n}$ and one knows that $Y \in F$ if and only if the determinants of all $n \times n$ matrices contained in Y are 0. This shows that at least 2 non-proportional irreducible polynomial functions vanish on F, so that F cannot contain a hypersurface. In the remaining case ((4) with $l+1$ odd) the argument is similar.

Remark. In all the cases of 11.20, the orbit condition is satisfied. This is easily checked, except in case (f). There the relation with J-structures, to be discussed in the next section will establish the assertion (see 12.4 and 12.7).

Notes

A problem about Lie algebras related to case 1 of the problem of 11.4 was discussed by Kostant [20]. Case 1 was first treated (in characteristic 0) by Serre, using a method different from the one followed here. For the special case where the multiplicities of the 2 weights are relatively prime see [28]. The general case was discussed by him in (unpublished) lectures at the Collège de France in 1967–1968.

In 2.21 we encountered already a situation similar to the one of case 2 of 11.4, viz. of a connected reductive group G and a 1-dimensional subtorus of G such that S has only 2 non-zero weights in the adjoint representation of G, which are opposite. It is not difficult to classify such pairs (G, S) with G semisimple, by using the procedure of this section (observe that one then has an analogue of 11.10(i)). A similar situation occurs also in the theory of bounded symmetric domains.

§12. Strongly Simple J-structures

Let $\mathscr{S} = (V, j, e)$ be a J-structure, with structure group G. We say that $\mathscr{S}$ is *strongly simple* if the identity component $G°$ acts irreducibly in V. Recall that $\mathscr{S}$ is said to be simple if there are no non-trivial ideals, see 1.3.

12.1 Lemma. (i) *A strongly simple J-structure is simple;*
(ii) *If* $\operatorname{char}(K) \neq 2$ *simple J-structures are strongly simple.*

Let $\mathscr{S} = (V, j, e)$ be strongly simple. The radical of $\mathscr{S}$ is a characteristic ideal of S (see 9.12(i)). It follows that $\mathscr{S}$ is semisimple. Then 10.22 shows that $\mathscr{S}$ is simple. This proves (i).

If $\operatorname{char}(K) \neq 2$, it follows from 9.7(ii) that a $G°$-stable subspace of V is an ideal, whence (ii).

Let $\mathscr{S} = (V, j, e)$ be strongly simple, assume $\dim V > 1$. 10.21 implies that V contains a primitive idempotent element $a \neq 0, e$. We use the notations of 11.1. So G' is the identity component of the intersection $G \cap SL(V)$, S' is the 1-dimensional torus constructed in 11.1 and C is the 1-dimensional torus of nonzero scalar multiplications in V.

12.2 Proposition. (V, G', S') *is admissible (in the sense of* 11.4*).* S' *has three weights in* V *and* $G° = CG'$ *has an open orbit in* V *whose complement contains a hypersurface.*

That S' has three weights in V follows from what was said in 11.1. It is then clear that (V, G', S') is admissible. Let N be the norm of $\mathscr{S}$, see 1.4. By 1.13, the orbit $G° \cdot e$ is open. Its complement contains the set of all points where N vanishes (by 1.5). This implies the last statement.

The results of §11 can now be invoked to give the classification of strongly simple J-structures. But before doing this we first establish some general facts, which we shall need. For N, σ and invariant bilinear forms see §1.

12.3 Proposition. *Let* $\mathscr{S} = (V, j, e)$ *be a strongly simple J-structure.*
(i) *The norm* N *of* $\mathscr{S}$ *is a irreducible polynomial function;*
(ii) *The standard symmetric bilinear form* σ *is nondegenerate, every invariant bilinear form on* $V \times V$ *is a multiple of* σ;

§12. Strongly Simple J-structures

(iii) *The structure group G of $\mathscr{S}$ is the group of all nonsingular linear transformations of V which leave the norm N invariant up to a nonzero scalar factor;*

(iv) *j is uniquely determined by N and e.*

Let B be a nonzero invariant form on $V \times V$, in the sense of 1.22. Such a form exists by 1.24. Put $R = \{x \in V \mid B(x, V) = 0\}$. Clearly R is a G°-stable subspace of V. $\mathscr{S}$ being strongly simple, we must have $R = \{0\}$. Hence B is nondegenerate. If B_1 is another invariant bilinear form there exists $a \in K$ such that $B_1 - aB$ is degenerate, hence $B_1 = aB$.

Next let F be an irreducible factor of N, assume that $F(e) = 1$. Let B be as in 1.24, it is nondegenerate by what we just established. Let $(e_i)_{1 \le i \le n}$ be a basis of V, let (e_i') be the dual basis with respect to B (i.e. $B(e_i, e_j') = \delta_{ij}$). It then follows from §1, (14) and §1, (21) that

$$jx = \sum_{i=1}^{n} F(x)^{-1}(dF)_x(e_i) \, e_i',$$

if x is invertible. But this shows that the denominator N of j must divide F, which can only be if $F = N$. This establishes (i).

If we take $F = N$ in 1.24, we see from §1, (11) that the form B is just the standard bilinear form σ. (ii) now follows from what was already established.

Let $g \in G$. 1.5 shows that there exists $a \in K^*$ such that

(1) $\qquad N(gx) = aN(x) \qquad (x \in V).$

Conversely, suppose that $g \in GL(V)$ satisfies (1). Denote by $(g')^{-1}$ the transposed of g with respect to the nondegenerate symmetric bilinear form σ. It follows from §1, (11) that we have

$$\sigma(j(gx), gy) = N(gx)^{-1}(dN)_{gx}(gy),$$

which equals $\sigma(jx, y)$, as follows from (1). Hence

$$\sigma((g')^{-1} j(gx), y) = \sigma(jx, y),$$

if x is invertible, for all $y \in V$. The nondegeneracy of σ shows that g is in the structure group G of S. This proves (iii).

By 1.18, N and e determine σ. (iv) now follows from §1, (11) and the nondegeneracy of σ.

12.4 Corollary. *Let $\mathscr{S}$ be as in 12.3, let $g \mapsto g'$ be the standard automorphism of the structure group G.*

(i) *For any rational character c of G we have $c(g') = c(g)^{-1}$;*

(ii) *The semi-invariants of $\mathscr{S}$ are the scalar multiples of the powers of N;*

(iii) *The complement of $G^\circ \cdot e$ contains only one irreducible hypersurface, namely $H = \{x \in V \mid N(x) = 0\}$.*

(Semi-invariants were defined in §1, just before 1.13.) We may assume $\dim V > 1$ (otherwise the assertions are trivial). Let $G^\circ = CG'$, with G' semisimple, C being the torus of scalar multiplications (see 12.2). Let $g \in G'$, $h \in C$. Since the semisimple group G' equals its commutator group, we have $c(g\,h) = c(h)$, for any rational character c. The definition of the standard automorphism (see 1.3) shows that $h' = h^{-1}$. Hence $c((g\,h)') = c(h') = c(h)^{-1} = c(g\,h)^{-1}$, proving (i).

Let F be a semi-invariant of $\mathscr{S}$. As in the proof of 1.12 one sees that the irreducible factors of F are also semi-invariants. So we may assume F to be irreducible, we may also assume that $F(e) = 1$.

Let c be the rational character of G defined by §1, (12). Then $c(g) = F(g\,e)$ and $F(j(g\,e)) = F(g' \cdot e) = c(g') = F(g\,e)^{-1}$, by (i). Hence $F(j\,x) = F(x)^{-1}$ if x is invertible. It follows that F divides a power of N, hence $F = N$ by 12.3(i). This establishes (ii).

The complement $S = V - G^\circ \cdot e$ is closed in V, hence a union of irreducible closed sets. The connected linear algebraic group G° permutes the (finitely many) irreducible components of S, hence stabilizes each of them. Let $H \subset S$ be an irreducible hypersurface, it is an irreducible component of S. We have that $G^\circ \cdot H = H$. There exists an irreducible polynomial function F such that $H = \{x \in V \mid F(x) = 0\}$, which is unique up to nonzero scalars. It follows that F is a semi-invariant of $\mathscr{S}$, hence by (ii) F is a scalar multiple of N, which proves (iii).

We can now prove that a strongly simple J-structure is completely determined by the action of its structure group.

12.5 Theorem. *Let $\mathscr{S} = (V, j, e)$ and $\mathscr{S}' = (V', j', e')$ be two strongly simple J-structures, with structure groups G and G', respectively. Let ϕ be a linear isomorphism of V onto V', such that $g \mapsto \phi \circ g \circ \phi^{-1}$ induces an isomorphism of G° onto $(G')^\circ$. Then $\mathscr{S}$ and $\mathscr{S}'$ are isomorphic.*

Let $\phi\,e = f$. Since the G°-orbit of e in V is open by 1.13, it follows that $(G')^\circ \cdot f$ is open in V'. Hence f lies in the unique open orbit $(G')^\circ \cdot e'$ of $(G')^\circ$ in V'. Let $h \in G'$ be such that $hf = e'$. Replacing ϕ by $h \circ \phi$, we are in the situation that we have $\phi\,e = e'$.

Let N and N' denote the norms of $\mathscr{S}$ and $\mathscr{S}'$. From 12.4(iii) we conclude that there exists $a \in K^*$ such that $N'(\phi\,x) = a\,N(x)$ ($x \in V$). Moreover, from $\phi\,e = e'$ it follows that $a = 1$. But then 1.18 implies that, σ and σ' denoting the standard symmetric bilinear forms,

$$\sigma'(\phi\,x, \phi\,y) = \sigma(x, y).$$

By 12.3(ii) σ and σ' are nondegenerate. Now §1, (11) implies that $j' \circ \phi = \phi \circ j$, which shows that ϕ is an isomorphism of J-structures.

§12. Strongly Simple J-structures

12.6 Classification of strongly simple J-structures. We now come to the main object of this section, viz. the classification of strongly simple J-structures.

We list below the special J-structures introduced in §2 and §5, together with the dimension n of their underlying vector space and the degree d of their norm (see §2 and §5). In the last column we have indicated when the corresponding J-structure is strongly simple. These results follow from what was established in 3.8 and just before 5.15.

J-structure	n	d	Strongly simple
$\mathcal{M}_r$	r^2	r	$r \geq 1$
$\mathcal{S}_r$	$\frac{1}{2}r(r+1)$	r	$r \geq 1$, $\mathrm{char}(K) \neq 2$
$\mathcal{A}_r$	$r(2r-1)$	r	$r \geq 1$
$\mathcal{O}_{2,r}$ (char $K \neq 2$)	r	2	$r \geq 3$
$\mathcal{O}'_{2,r}$ (char $K = 2$)	$2r$	2	$r \geq 2$
$\mathcal{E}_3$	27	3	yes

The next theorem shows that these are all the strongly simple J-structures.

12.7 Theorem. *Let $\mathcal{S}$ be a strongly simple J-structure.*
(i) *If $\mathrm{char}(K) \neq 2$, then $\mathcal{S}$ is isomorphic to one of the J-structures $\mathcal{M}_r$ $(r \geq 1)$, $\mathcal{S}_r$ $(r \geq 2)$, $\mathcal{A}_r$ $(r \geq 3)$, $\mathcal{O}_{2,r}$ $(r \geq 5)$, $\mathcal{E}_3$;*
(ii) *If $\mathrm{char}(K) = 2$, then $\mathcal{S}$ is isomorphic to one of the J-structures $\mathcal{M}_r$ $(r \geq 1)$, $\mathcal{A}_r$ $(r \geq 3)$, $\mathcal{O}'_{2,r}$ $(r \geq 3)$, $\mathcal{E}_3$;*
(iii) *No two of the J-structures listed in (i) and (ii), respectively, are isomorphic.*

First observe that (iii) is a consequence of the fact that the J-structures listed in (i) and (ii) can be distinguished by the values of n and d. Also observe the following isomorphisms:

$$\mathcal{M}_1 \simeq \mathcal{S}_1, \quad \mathcal{M}_1 \simeq \mathcal{A}_1,$$
$$\mathcal{O}_{2,3} \simeq \mathcal{S}_2, \quad \mathcal{O}_{2,4} \simeq \mathcal{M}_2, \quad \mathcal{O}_{2,6} \simeq \mathcal{A}_2 \quad (\text{if } \mathrm{char}(K) \neq 2),$$
$$\mathcal{O}'_{2,2} \simeq \mathcal{M}_2, \quad \mathcal{O}'_{2,3} \simeq \mathcal{A}_2, \quad (\text{if } \mathrm{char}(K) = 2).$$

The first isomorphisms are trivial, the others follows from the discussion of 5.1 (taking into account the properties of $\mathcal{M}_2, \mathcal{A}_2, \mathcal{S}_2$, discussed in §2). It follows that all strongly simple J-structures listed in 12.6 occur among the ones listed in 12.7(i) and (ii).

If $\dim V = 1$, we know by 2.19 that $\mathcal{S}$ is isomorphic to $\mathcal{M}_1$. So assume now $\dim V > 1$. We use the notations introduced just before 12.2. It follows from 12.2 that 11.20 is applicable, so that we have the possibilities of 11.20 for G' and V.

We now determine, for each of the J-structures listed in (i) and (ii) the corresponding case of 11.20. First let $\mathscr{S} = \mathscr{O}_{2,r}$ (char$(K) \neq 2, r \geq 2$). From 2.18 it follows that we have $G' = \mathbb{SO}_r$. But then we must have case (d) or (e) of 11.20, these being the only ones where a group $\mathbb{SO}_r$ occurs. Similarly, if char$(K) = 2$, and if $\mathscr{S} = \mathscr{O}'_{2,r}$, we have case (d) of 11.20.

Next a consideration of dimensions shows that $\mathscr{E}_3$ corresponds to case (f) of 11.20 (which is the only one where the dimension 27 arises).

It remains to identify $\mathscr{M}_r$, $\mathscr{S}_r$ and $\mathscr{A}_r$. This can be done by distinguishing them by comparing the different values of n and d. Comparing with the results of 11.20, taking into account the isomorphisms in low dimensions stated above, one concludes that $\mathscr{M}_r$ corresponds to case (a), $\mathscr{S}_r$ (char$(K) \neq 2$, $r \geq 2$) to case (b) and $\mathscr{A}_r$ ($r \geq 3$) to case (c) of 11.20. But now we see that all cases listed in 11.20 come from J-structures.

Moreover 12.5 shows that to each case there corresponds only one strongly simple J-structure (up to isomorphism). This implies that (i) and (ii) give all strongly simple J-structures if char$(K) \neq 2$ or char$(K) = 2$, respectively.

12.1 shows that the preceding theorem gives the full classification of simple J-structures if char$(K) \neq 2$. By what was remarked in the first paragraph of 9.8 we then also obtain the classification of simple Jordan algebras if char$(K) \neq 2$. In characteristic 2 we still have to deal with simple but not strongly simple J-structures. This will be done in the next section (it will be seen that such J-structures are those of types $\mathscr{S}_r$ and $\mathscr{O}''_{2,r}$).

12.8 Simple associative algebras. As another application of the results of Section 11, we shall indicate how to obtain Wedderburn's theorem on the structure of simple associative algebras, over the algebraically closed field K.

Let A be a finite dimensional associative algebra over K, with identity element. Recall that A is said to be simple, if $\{0\}$ is the only proper twosided ideal in A.

12.9 Theorem. *Let A be a simple associative algebra. Then A is isomorphic to a matrix algebra $\mathbb{M}_r$.*

We may assume that dim $A > 1$, otherwise the assertion is trivial. Let $\mathscr{J}(A)$ be the J-structure defined by A (see 2.1). Denote by A^* the group of invertible elements of A. A^* is a connected linear algebraic group. Let $H \subset GL(A)$ be the group of linear transformations of the form $x \mapsto a x b^{-1}$ ($a, b \in A^*$), this is a connected linear algebraic group.

If I is an H-stable subspace of A, then $A^* I A^* \subset I$, hence $A I A \subset I$, so that I is a twosided ideal. The simplicity of A now implies that H acts irreducibly in A. Since we know by 2.1 that H is contained in the structure group of $\mathscr{J}(A)$, it follows that $\mathscr{J}(A)$ is strongly simple, hence isomorphic

§ 12. Strongly Simple J-structures

to one of the J-structures of 12.7. However we shall not apply 12.7 here, but use instead the results of Section 11 directly. We only use the strong simplicity of $\mathscr{J}(A)$ to obtain the existence of a primitive idempotent $a \neq e$ in A (of course, this can also be deduced directly from the simplicity of the associative algebra A).

Let S_a be the 1-dimensional torus of 10.2. Using 2.2 one sees that $S_a \subset H$. Let H', S' be the identity components of the intersections of H and S with $SL(A)$, respectively. It follows as in 12.2 that (A, H', S') is an admissible triple in the sense of 11.4. Moreover, H' is not quasi-simple, and H has an open orbit in A whose complement contains a hypersurface. It follows that we must be in case (a) of 11.20.

Hence there exists a linear isomorphism $\phi: A \to \mathbb{M}_r$ and isomorphisms of linear algebraic groups $\psi_i: A^* \to \mathbb{GL}_r$ ($i=1, 2$) such that $\phi(e) = e$ and that

$$\phi(a\,x\,b^{-1}) = \psi_1(a)\,\phi(x)\,\psi_2(b)^{-1} \qquad (a, b \in A^*, x \in A).$$

Putting $x = b = e$, we obtain $\psi_1(a) = \phi(a)$, likewise $\psi_2(a) = \phi(a)$. It follows that ϕ is an isomorphism of A onto $\mathbb{M}_r$.

Notes

Using § 6 one obtains from 12.7 Albert's classification of simple Jordan algebras over an algebraically closed field of characteristic $\neq 2$. We leave it to the reader to work out the details. For a treatment of this subject which is more elementary than the one given here see [8, X] or [14, Chapter V]. We shall discuss the case of an arbitrary groundfield in § 15.
It would be interesting to have a group theoretical determination of simple Artin rings, in the manner of 12.9.

§13. Simple J-structures

In this section we shall complete the determination of the simple J-structures by determining the simple ones which are not strongly simple. $\mathscr{S} = (V, j, e)$ denotes a J-structure, with structure group G. First we establish some results about G°-stable subspaces of V.

13.1 Lemma. *Let $W \in V$ be a G°-stable subspace of V.*
(i) *We have $P(W, W) V \subset W$, $P(V) W \subset W$;*
(ii) *$W' = \{w \in W \mid P(w) V \in W\}$ is an ideal of $\mathscr{S}$.*

That $P(V) W \subset W$ is established as in the proof of 9.7. If $w_1, w_2 \in W$, $g \in G$ then 1.16(i) and §3, (10) imply that

$$P(w_1, w_2)(g\,e) = (g')^{-1} P(g' \cdot w_1, g' \cdot w_2)\, e = (g')^{-1} P(g' \cdot w_1, e)\, g' \cdot w_2$$

($g \mapsto g'$ denoting the standard automorphism). From $P(V) W \subset W$ it follows that $P(V, V) W \subset W$. The preceding formula then shows that $P(W, W)(G^\circ \cdot e) \subset W$, whence $P(W, W) V \subset W$ (using 1.13).

To prove (ii), first notice that (i) implies that W' is a subspace of W. By 9.6 it suffices to show that $P(V) W' \subset W'$ and $P(W') V \subset W'$. Let $v \in V$, $w \in W'$. Then we have by §3, (6)

$$P\big(P(v)\,w\big)\,V = P(v)\,P(w)\,P(v)\,V \subset P(V)\,W \subset W$$

and

$$P\big(P(w)\,v\big)\,V = P(w)\,P(v)\,P(w)\,V \subset P(w)\,P(v)\,W \subset W,$$

establishing the required results.

Now assume that $\mathscr{S}$ is simple, but not strongly simple. It follows from 9.7(ii) that we must have $\operatorname{char}(K) = 2$.

13.2 Lemma. *Let $W \subset V$ be a proper nonzero subspace which is G°-stable.*
(i) *If $w \in W$, $P(w) V \subset W$ then $w = 0$;*
(ii) *No primitive idempotent of $\mathscr{S}$ is contained in W.*

(i) is a consequence of 13.1(ii). (ii) follows from (i) if one bears in mind that if a is a primitive idempotent we have $P(a) V = K a$, by 10.3(ii) and 10.21.

We next deal with a particular case.

§13. Simple J-structures

13.3 Proposition. *Let $\mathscr{S}$ be simple with $\dim V > 1$. Assume that there exists a nonzero $x \in V$ such that $G°(Kx) = Kx$. Then the degree of $\mathscr{S}$ is 2 and $\mathscr{S}$ is a J-structure defined by a nondegenerate, defective quadratic form.*

The J-structure defined by a quadratic form Q was introduced in §2. Recall that Q is said to be defective if the associated bilinear form $Q(\ ,\ )$ is degenerate.

Let x be as stated. Then 1.16(i) shows that $P(x)V$ is a $G°$-stable subspace. Moreover $P(P(x)V)V \subset P(x)V$. 9.7(i) now shows that $P(x)V$ is an ideal. If it were the zero ideal, we had $(gx)^2 = P(gx)e = 0$ for all $g \in G$ and x was a radical element of $\mathscr{S}$ (see 9.10). Since $x \neq 0$ this contradicts the semisimplicity of $\mathscr{S}$. Consequently we have $P(x)V = V$, hence $P(x)$ is invertible.

Assume that $S \in G°$ is a torus. There exists a character a of S such that
$$s x = s^a x \quad (s \in S).$$
Let y be a nonzero weight vector for S in V, with weight b.

From 1.16(i) it follows that
$$s(P(x)y) = s^{2a-b}(P(x)y).$$
Since $P(x)y \neq 0$ we see that $2a - b$ is also a weight of S in V.

Now let $a_1, \ldots, a_h$ be a set of mutually orthogonal idempotents with sum e. Applying the preceding remark to the torus
$$S = \{P(t_1 a_1 + \cdots + t_h a_h) \mid t_i \in K^*\}$$
of 10.26(iii) we conclude from the form of the weights of S in V (see 10.27) that we must have $h \leq 2$.

Let d be the degree of the norm N. Choose $a \in V$ such that it has the property of 8.3 and that none of its powers is 0 or e (using 8.3 and 9.17 one sees that the set of these a is open and nonempty).

Let $a = a_s + a_n$ be the Jordan decomposition of 8.9. We can write a_s as a linear combination of primitive idempotents (by 9.14 and 10.19) and the number of these is 2, by what we established in a preceding paragraph. Our choice of a now shows that there is a primitive idempotent b such that $e - b$ is also a primitive idempotent and that $a_s \in Kb + K(e - b)$.

b is a linear combination of powers of a_s, hence also of powers of a (by 8.9). It then follows from §3,(20) that $P(b, e)a_n = 0$, whence $a_n \in Kb + K(e - b)$, by 10.3(ii) and 10.21. This implies that $a_n = 0$. Applying 10.29 we obtain that $d = 2$.

Then we have seen in 5.1 that $\mathscr{S}$ is a J-structure defined by a quadratic form. Since $\mathscr{S}$ is simple, it follows from 9.11 that the quadratic form must be nondegenerate. What was said in 3.8 then implies that the form is defective.

Let again $\mathscr{S}$ be simple but not strongly simple. Assume that $W \neq V$ is a maximal $G°$-stable subspace. Then $G°$ acts irreducibly in V/W and

$W \neq \{0\}$. Let a be a primitive idempotent of $\mathscr{S}$. By 13.2(ii) we know that $a \notin W$. Let S_a be the 2-dimensional torus defined in 10.1, let $V = V_a \oplus V'_a \oplus V_{e-a}$ be the Peirce decomposition defined by a.

The following lemma, which is proved by using the results of section 11, contains the crucial step in the classification.

13.4 Lemma. *Assume that W does not contain a $G°$-stable subspace of dimension 1. Then $V'_a \subset W$. If $\dim V/W = r$, we have $\dim W \geq \frac{1}{2} r(r-1)$.*

Let $W_1 \subset W$ be a nonzero $G°$-stable subspace of W in which $G°$ acts irreducibly. We have $\dim W_1 > 1$. Let U be the unipotent radical of $G°$, put $H = G°/U$. Denote by S the canonical image of S_a in H. H acts irreducibly in V/W and W_1, moreover S has at most 3 weights in the first representation and at most 2 in the second one.

In fact, the weight vector a of S_a determines a weight vector for S with multiplicity 1 in V/W and the corresponding weight of S cannot occur in W_1.

H being reductive, there exist quasi-simple linear algebraic groups $H_1, \ldots, H_s$, a torus A and a central isogeny

$$\pi \colon H' = H_1 \times \cdots \times H_s \times A \to H.$$

Let ρ and σ denote the irreducible representations of H' in V/W and W_1, obtained by composing with π the corresponding representations of H. If $\dim V/W = 1$ we must have $V'_a \subset W$ and the lemma is proved.

If $\dim V/W > 1$ then S has at least 2 distinct weights in V/W, hence S cannot be a central torus of H. It follows that S has 2 distinct weights in W_1 (if not, all elements of $G°$ had to act as scalar multiplications in W_1, and V had a $G°$-stable subspace of dimension 1).

11.6 and 11.7 now imply that we may assume there exists a 1-dimensional torus S_1 in H_1 such that $\sigma(S) = K^* \cdot \sigma(S_1 \times \{1\})$, so that S comes from the quasi-simple group H_1. But it then follows from 11.7 that ρ is trivial on $\{1\} \times H_2 \times \cdots \times H_s \times A$.

Identify H_1 with a subgroup of H'. Then $(V/W, \rho(H_1), \rho(S_1))$ is an admissible triple in case 1 with a weight vector of multiplicity 1, or in case 2 (in the sense of 11.4). Also, there is a decomposition $W_1 = W'_1 \otimes W''_1$ and an irreducible representation σ' of H_1 in W'_1 such that $(W'_1, \sigma'(H_1), \sigma'(S_1))$ is an admissible triple in case 1. Moreover, ρ and σ' are inequivalent irreducible representations of H_1, since they behave differently on S_1.

We now have the following situation: H_1 is a quasi-simple linear algebraic group, S_1 a 1-dimensional torus in H_1, ρ and σ' two irreducible representations of H_1 in the spaces V/W and W'_1, respectively, such that $(V/W, \rho(H_1), \rho(S_1))$ is admissible in case 1 or 2 and that $(W'_1, \sigma'(H_1), \sigma'(S_1))$ is admissible in case 1. Moreover, if $(V/W, \rho(H_1), \rho(S_1))$ is in case 2, the 2 weights of S_1 in W'_1 occur among the 3 weights of S_1 in V/W. We shall

§ 13. Simple J-structures

apply the classification of admissible triples given in Section 11 to discuss this situation.

As in 11.5, let T be a maximal torus of H_1 containing S_1 and choose compatible orderings on the character groups $X(T)$ and $X(S_1)$. We use the notations of 11.16. From what we established there, it follows that the cases which arise can be described as follows.

Denote by ϖ and ϖ' the highest weights of the first and second representations (ρ and σ', respectively). The situation (B) of 11.14 cannot occur if the representation ρ is in case 2, hence there is a simple root α such that $S_1 = S_\alpha$ (notation of 11.14(i)).

(a) S_1 has 3 weights in V/W and 2 in W_1.

The cases are the following.

Type A_l: $\alpha = \alpha_2$, $\varpi = \varpi_2$, $\varpi' = \varpi_1$ or ϖ_l,

$\alpha = \alpha_{l-1}$, $\varpi = \varpi_{l-1}$, $\varpi' = \varpi_1$ or ϖ_l,

$\alpha = \alpha_1$, $\varpi = 2\varpi_1$, $\varpi' = \varpi_j$ $(1 < j < l)$,

$\alpha = \alpha_l$, $\varpi = 2\varpi_l$, $\varpi' = \varpi_j$ $(1 < j < l)$.

Type B_l: $\alpha = \alpha_1$, $\varpi = \varpi_1$, $\varpi' = \varpi_l$.

Type D_l: $\alpha = \alpha_1$, $\varpi = \varpi_1$, $\varpi' = \varpi_{l-1}$.

(From the discussion of 11.16 we also obtain some extra possibilities for types D_4 and D_5. However those in case D_4 lead to the transforms of the previous case by one of the outer automorphism of a group of type D_4, so do not lead to anything new. From the discussion in 11.17 it follows that the extra cases in type D_5 are impossible)

The last 2 possibilities in type A_l can be ruled out, since in characteristic 2 the torus S_1 has only 2 weights in the irreducible representations with highest weight $2\varpi_1$ and $2\varpi_l$ (see the discussion in 11.17).

The cases of types B_l and D_l are also impossible: in that case S_1 has a zero weight in the representation ρ, with multiplicity >1 (see 11.17), whereas σ' is a spin representation, where a weight 0 does not occur. It is also easily seen that it suffices to deal with the first case in type A_l: $\alpha = \alpha_2$, $\varpi = \varpi_2$, $\varpi' = \varpi_1$ or ϖ_l. We shall show that this case is impossible.

The map $W \times V \to V$ with $(w, v) \mapsto P(w)v$ defines (using 13.1(i)) a map $F: W_1 \times V/W \to V/W$ which is biadditive and linear in its second variable. Moreover $F(ax, y) = a^2 F(x, y)$ ($a \in K$). From 13.2(i) we see that $F \neq 0$.

The standard automorphism of G induces an automorphism $h \mapsto h'$ of H_1 and from 1.16(i) we see that

$$F(\sigma(h)x, \rho(h')y) = \rho(h)F(x, y) \quad (x \in W_1, y \in V/W, h \in H_1).$$

Let $\sigma^{[2]}$ be the transform of σ by the Frobenius automorphism, let ρ' be the representation $h \mapsto \rho(h')$ of H_1. The last formula then shows that ρ is a quotient representation of the tensor product $\sigma^{[2]} \otimes \rho'$. In the case we are dealing with, a consideration of the weights of T in the various representa-

tions shows that this is impossible. So we conclude that case (a) cannot occur. Hence we must have the other case:
(b) S_1 has 2 weights in V/W and 2 in W_1.

One possibility is the one we already mentioned in (a):

$$\alpha = \alpha_1, \varpi = 2\varpi_1, \varpi' = \varpi_j \quad (1 < j < l) \quad (\text{resp. } \alpha = \alpha_l, \varpi = 2\varpi_l, \varpi' = \varpi_j).$$

In that case we do indeed have $V'_a = W$. Moreover a comparison of weights of $2\varpi_1$ and ϖ_j (using the results of 11.17) shows that the only possibility is $j=2$ (resp. $j=l-1$). The representation in V/W has dimension $r=l+1$, that in W_1 is the second exterior power representation, of dimension $\frac{1}{2}r(r-1)$. This establishes the assertion in that case.

There might be another possibility in type A_l: $\alpha = \alpha_1, \varpi = \varpi_1, \varpi' = \varpi_j$ $(1 < j < l)$. However the discussion of 11.17 shows that then the representation in V/W and W_1 have no common weight, hence this case is impossible. This concludes the proof of 13.4.

We can now determine the simple J-structures which are not strongly simple.

13.5 Theorem. *Suppose that $\mathscr{S}$ is simple but not strongly simple. Then* char$(K)=2$ *and $\mathscr{S}$ is isomorphic to $\mathcal{O}''_{2,r}$ $(r \geq 2)$ or $\mathscr{S}_r$ $(r \geq 3)$.*

We have already observed that K must have characteristic 2. If V contains a 1-dimensional $G°$-stable subspace then either dim $V=1$ (which is impossible) or $\mathscr{S}$ is isomorphic to $\mathcal{O}''_{2,r}$ by 13.3. So we may assume that we are in the situation discussed in 13.4 and we have to prove that $\mathscr{S}$ is isomorphic to $\mathscr{S}_r$ $(r \geq 3)$.

Let W be as in 13.4. Let $a_1, \ldots, a_r$ be a set of mutually orthogonal primitive idempotents of $\mathscr{S}$ with sum e. Denote by $S \subset G°$ the r-dimensional torus consisting of the linear transformations $P(t_1 a_1 + \cdots + t_r a_r)$ $(t_i \in K^*)$, let $V = \bigoplus_{1 \leq s \leq t \leq r} V_{st}$ be the decomposition given in 10.27.

We have $V_{ss} = K a_s$ by 10.21, moreover 13.4 shows that $V_{st} \subset W$ $(s \neq t)$, hence V/W is spanned by the images of the a_s (observe that by 13.2(ii) no a_s lies in W). Let $v \in V_{st}$ $(s \neq t)$, $w \in V_{kl}$. By considering the action of S one sees that $P(v) w \in W$, unless $V_{kl} = V_{ss}$ or V_{tt} and that $P(v) a_s = Q_s(v) a_t$, where Q_s is a quadratic form on V_{st}.

By §3, (6) it follows that

$$Q_s^2(v) a_t = P(P(v) a_s) a_t = P(v) P(a_s) P(v) a_t = Q_s(v) Q_t(v) a_t,$$

from which it follows that $Q_s(v) = Q_t(v)$ $(v \in V_{st})$. This implies that

$$P(v) W \subset W + K Q_s(v)(a_s + a_t).$$

§ 13. Simple J-structures

13.2(i) then shows that $Q_s(v)=0$ implies $v=0$, which gives that $\dim V_{st} \leq 1$. Since $\dim W \geq \frac{1}{2}r(r-1)$ by 13.4, we must have $\dim V_{st}=1$, $\dim W=\frac{1}{2}r(r-1)$.

Let a_{st} be a nonzero element in V_{st} with $a_{st}^2 = a_s + a_t$. Put $a_{ss}=a_s$. We define a linear isomorphism ϕ of V onto the space $\mathbb{S}_r$ of symmetric $r \times r$ matrices by

$$\phi\Big(\sum_{1 \leq s \leq t \leq r} c_{st} a_{st}\Big) = \sum_{s=1}^{r} c_{ss} e_{ss} + \sum_{1 \leq s < t \leq r} c_{st}(e_{st}+e_{ts}),$$

where (e_{st}) is the canonical basis of the matrix algebra $\mathbb{M}_r$.

We shall prove that ϕ is an isomorphism of J-structures of $\mathscr{S}$ onto $\mathscr{S}_r$. It is clear that $\phi(e)$ is the unit matrix. Using 1.16(iii) one sees then that ϕ will be an isomorphism if we have the following relation for the quadratic maps of $\mathscr{S}$ and $\mathscr{S}_r$:

(1) $$\phi \circ P_{\mathscr{S}}(x) = P_{\mathscr{S}_r}(\phi(x)) \circ \phi \qquad (x \in V).$$

To prove (1) we determine the elements $P(a_{st})\,a_{kl}$ and $P(a_{st}, a_{kl})\,a_{mn}$ and we show that their values are consonant with (1). By considering the action of the torus S one sees that the nonzero elements of this form occur among the following ones:

$$P(a_{st})\,a_{st}, \quad P(a_{st})\,a_{ss}, \quad P(a_{st}, a_{sk})\,a_{st}, \quad P(a_{st}, a_{kl})\,a_{st}, \quad P(a_{st}, a_{kl})\,a_{sk}$$

(that the others are 0 is in accordance with (1)).

We shall prove, as an example of the determination of the above elements, the following formulas:

(a) $\qquad P(a_{st})\,a_{ss}=a_{tt} \qquad (s \neq t)$,

(b) $\qquad P(a_{st})\,a_{st}=a_{st} \qquad (s \neq t)$,

(c) $\qquad P(a_{st}, a_{sk})\,a_{st}=a_{sk} \qquad (s, t, k \text{ distinct})$.

The proof of these formulas will show how to proceed, and the reader will have no difficulty in completing the argument.

First remark that, by using the action of S, one sees that the formulas to be proved are true op to a scalar factor. Since $a_{st}^2=a_s+a_t$, we have $P(a_{st})\,e=P(a_{st})(a_s+a_t)=a_s+a_t$, whence (a).

Let $P(a_{st})\,a_{st}=c\,a_{st}$. We have

$$c^2(a_s+a_t) = P(P(a_{st})\,a_{st})\,e = P(a_{st})^3\,e = P(a_{st})^3(a_s+a_t) = a_s+a_t,$$

by (a), hence $c=1$. This proves (b).

Put $P(a_{st}, a_{sk})\,a_{st}=c\,a_{sk}$. Applying §3, (7) with $x=a_{st}$, $y=a_{sk}$, $z=a_{st}$, we obtain

$$c^2 P(a_{sk}) + P(P(a_{st})\,a_{st}, P(a_{sk})\,a_{st}) = P(a_{st})^2\,P(a_{sk})$$
$$+ P(a_{sk})\,P(a_{st})^2 + P(a_{st}, a_{sk})\,P(a_{st})\,P(a_{st}, a_{sk}).$$

Since $P(a_{sk})a_{st}=0$, this gives

$$c^2 a_{sk}^2 = P(a_{st})^2 a_{sk}^2 + P(a_{sk}) a_{st}^4 + P(a_{st}, a_{sk}) P(a_{st}) P(a_{st}, a_{sk}) e.$$

Inserting the values of a_{sk}^2 and a_{st}^4 and using (a), we see that

$$c^2(a_s+a_k) = a_s+a_k + P(a_{st}, a_{sk}) P(a_{st}) P(a_{st}, a_{sk}) e.$$

Now the third term in the right-hand side lies in W, whence we conclude that it must be 0. It follows that $c=1$.

13.6 Theorem. (i) *A simple J-structure is isomorphic to one of the J-structures* $\mathcal{M}_r$ $(r \geq 1)$, $\mathcal{S}_r$ $(r \geq 2)$, $\mathcal{A}_r$ $(r \geq 3)$, $\mathcal{E}_3$, $\mathcal{O}_{2,r}$ (char$(K) \neq 2$, $r \geq 5$), $\mathcal{O}'_{2,r}$ (char$(K)=2$, $r \geq 3$), $\mathcal{O}''_{2,r}$ (char$(K)=2$, $r \geq 2$).
(ii) *No two of the J-structures listed in* (i) *are isomorphic*.

(i) follows from 13.5 and 12.7. The proof of (ii) is like that of 12.7(iii).

13.7 Corollary. *Let* char$(K)=2$. *Let* $\mathcal{S}$ *be a simple J-structure. Then* $\mathcal{S}$ *is separable.*

This follows from 13.6 and 7.9.

We now discuss some characterizations of simple and strongly simple J-structures, which are consequences of the classification theorems. We denote by N and σ the norm and standard symmetric bilinear form of the J-structure $\mathcal{S}$.

13.8 Proposition. (i) $\mathcal{S}$ *is simple if and only if N is a nondegenerate irreducible polynomial function;*
(ii) $\mathcal{S}$ *is strongly simple if and only if $\mathcal{S}$ is simple and σ is nondegenerate.*

If $\mathcal{S}$ is simple, then the classification theorem 13.6 together with the determination of norms in §2 and §5, shows that N is nondegenerate and irreducible.

Assume that $\mathcal{S}$ is not simple, let $I \neq V$ be a nonzero ideal. Denote by J a complementary subspace of I in V. Put $jx=j_1 x+j_2 x$, where $j_1(x) \in I$, $j_2(x) \in J$ (x invertible). j_1 and j_2 are rational maps of V into I, J respectively.

The definition of ideals (see 1.3) shows that $j_2(x+y)=j_2(x)$ if x is invertible and $y \in I$. Let N_i be a denominator of j_i ($i=1, 2$). The polynomial functions N_i are divisors of N. We have $N_2(x+y)=N_2(x)$ ($x \in V$, $y \in I$). If N is irreducible we must have $N_2=cN$, but then N is degenerate. It follows that if $\mathcal{S}$ is not simple, N is degenerate or reducible. This concludes the proof of (i).

If $\mathcal{S}$ is simple but not strongly simple, then 13.5 and the results of §2 show that σ is degenerate. (ii) now follows from (i), using 3.7.

§ 13. Simple J-structures

13.9 Proposition. *$\mathscr{S}$ is a direct sum of strongly simple J-structures if and only if σ is nondegenerate.*

If $\mathscr{S}$ is a direct sum of strongly simple J-structures then it follows from 13.8(ii) that σ is nondegenerate.

Conversely, if σ is nondegenerate, the definition of σ (see 1.11) shows that N must be nondegenerate. Then 9.11 shows that $\mathscr{S}$ is semisimple, hence a direct sum of simple J-structures by 10.22. By what was said in 1.25, each of these has a nondegenerate standard symmetric bilinear form, hence is strongly simple by 13.8(ii).

13.10 Corollary. *If $\operatorname{char}(K) \neq 2$, then $\mathscr{S}$ is semisimple if and only if σ is nondegenerate.*

If $\operatorname{char}(K) \neq 2$, simplicity and strong simplicity coincide. 13.10 now follows from 13.9 and 10.22.

Notes

The classification of simple J-structures given in 13.6 is closely related to the classification of simple quadratic Jordan algebras over an algebraically closed field, which is given in [15, p. 3.61]. Using 7.10(i) one sees that the latter result implies 13.6. Conversely, by 7.10(ii) we have that 13.6 is equivalent to the determination of the *separable* simple quadratic Jordan algebras. A posteriori, it follows from the classification that all simple quadratic Jordan algebras are separable. If a direct proof of this fact were available then 13.6 could be considered to be equivalent to the classification of simple quadratic Jordan algebras.

§14. The Structure Group of a Simple J-structure and the Related Lie Algebras

In this section we discuss in detail the structure groups of the various simple J-structures of the preceding sections. We also investigate the various Lie algebras introduced in §4. Before doing so we establish some auxiliary results about root systems and derivations of Lie algebras. Throughout, p denotes the characteristic of K.

14.1 A lemma on root systems. Let R be a finite dimensional vector space over $\mathbb{R}$, let Σ be a root system in V, in the sense of [7, p. 142], with Weyl group W. Let $(\,|\,)$ be a positive definite W-invariant scalar product in $R \times R$. If $\alpha \in \Sigma$ we denote by s_α the reflection in R defined by α. Let $\Delta \subset \Sigma$ be a set of linearly independent roots.

We attach to Δ a graph $\Gamma(\Delta)$ as follows. The vertices of $\Gamma(\Delta)$ are the elements of Δ; if $\alpha, \beta \in \Delta$ then $\{\alpha, \beta\}$ is an edge of $\Gamma(\Delta)$ if and only if s_α and s_β do not commute. Moreover let f be the map of the set of edges of $\Gamma(\Delta)$ into $\mathbb{Z}$ defined by the following rule. Let $\{\alpha, \beta\}$ be an edge, let $m_{\alpha\beta}$ be the order of $s_\alpha s_\beta$, let $\varepsilon_{\alpha\beta} = -\operatorname{sgn}(\alpha|\beta)$. Then

$$f(\{\alpha, \beta\}) = \varepsilon_{\alpha\beta} m_{\alpha\beta}.$$

Notice that f is independent of the choice of the scalar product.

We call the pair $(\Gamma(\Delta), f)$ the *Coxeter graph* defined by Δ. If Δ is a basis of Σ, then $(\Gamma(\Delta), f)$ is isomorphic to the Coxeter graph defined by Σ, introduced in [7, p. 189]. We need the following converse.

14.2 Lemma. *Assume that $(\Gamma(\Delta), f)$ is isomorphic to the Coxeter graph of Σ. Then Δ is a basis of Σ.*

Let $S = \{x \in R \,|\, (x|x) = 1\}$, let μ be a nonzero measure on the sphere S which is invariant under all linear isometries of R (with respect to the scalar product). Let C be a chamber in R relative to W, in the sense of [7, p. 73], let $\bar{C}$ be its closure in R. Then $F = \bar{C} \cap S$ is a fundamental domain in S for the action of W on S (as follows from [loc. cit., Théorème 2, p. 75]), from which one concludes that $\mu(S) = \mu(F)|W|$ ($|W|$ denotes the order of W).

§ 14. The Structure Group of a Simple J-structure

Let W_1 be the subgroup of W generated by the reflections s_α with $\alpha \in \Delta$. Let $D = \{x \in R | (x|\alpha) > 0 \text{ for all } \alpha \in \Delta\}$. Clearly D contains a chamber C_1 of V relative to W_1. Put $\bar{D} \cap S = G$, $\bar{C}_1 \cap S = F_1$. From the assumption that $(\Gamma(\Delta), f)$ is isomorphic to the Coxeter graph defined by Σ it follows that there exists an isometry t of R with $t(C) = D$, whence $t(G) = F$. Hence $\mu(F) = \mu(G)$, consequently $\mu(F_1) \leq \mu(F)$.
Since

$$\mu(S) = \mu(F_1) \cdot |W_1| \leq \mu(F) \cdot |W| \leq \mu(S),$$

we find that $W = W_1$, $F = F_1$, $C = C_1$. This implies our assertion.

14.3 Some results on derivations of Lie algebras. Let G be a reductive linear algebraic group, let T be a maximal torus of G. Let Σ be the root system of G with respect to T. Denote by X the character group of T and by Y the subgroup of X generated by Σ. We denote by $L(G)$, $L(T)$, ... the Lie algebras of G, T, ...

Our aim is to prove a result about derivations of $L(G)$ (viz. 14.5). We need an auxiliary result about tori, which we first establish. If a is a character of a torus S, then da denotes its differential, which is a linear function on $L(S)$.

14.4 Lemma. *Let S be a torus with character group Z. Let Z_1 be a subgroup of Z such that Z/Z_1 has no p-torsion if $p > 0$. Let $U = \bigcap_{a \in Z_1} \mathrm{Ker}\, a$. Then U is an algebraic subgroup of S with Lie algebra*

$$L(U) = \bigcap_{a \in Z_1} \mathrm{Ker}\,(da).$$

That U is an algebraic subgroup is clear. Let

$$Z_1 \supset Z_2 \supset \cdots \supset Z_k \supset Z_{k+1} = \{0\}$$

be a sequence of subgroups of Z_1 such that Z_i/Z_{i+1} is cyclic. Clearly Z/Z_i has no p-torsion. Let $U_i = \bigcap_{a \in Z_i} \mathrm{Ker}\, a$. Then

$$U_1 \subset U_2 \subset \cdots \subset U_k \subset U_{k+1} = S.$$

We use induction on k. Suppose $k > 1$ and that the assertion has already been established for smaller values of k. Then we may assume

$$L(U_k) = \mathrm{Ker}\,(da_k)$$

where a_k is a suitable generator of U_k. The induction assumption, applied to the identity component of U_k (which is a torus) then establishes the result.

Hence we are reduced to proving the statement in the case $k=1$. Let a be a generator of Z_1. The absence of p-torsion in Z/Z_1 implies that the differential da is a nonzero linear function on $L(S)$. Its kernel clearly contains $L(U)$. Since $\dim L(U) = \dim U = \dim S - 1 = \dim \operatorname{Ker}(da)$, we have $L(U) = \operatorname{Ker}(da)$.

Let now G, T, X, Y be as stated in 14.3. Let C be the center of G.

14.5 Proposition. *Assume the following:*
  (i) X/Y *has no p-torsion (if $p > 0$);*
  (ii) *If $p=2$ then Σ has no simple component of type A_1;*
  (iii) *If $p=2$ then all simple components of Σ have equal root lengths (i.e. are of one of the types A_l, D_l, E_6, E_7, E_8);*
  (iv) *If $p=3$ then no simple component of Σ is of type G_2.*

Let D be a derivation of $L(G)$ which commutes with all $\operatorname{Ad}(t)$ ($t \in T$). Then D is a sum of an inner derivation of $L(G)$ (i.e. of the form $DX = [A, X]$ with $A \in L(G)$) and a derivation of $L(G)$ into $L(C)$.

14.6 Corollary. *If moreover C is finite then D is an inner derivation.*

14.6 is an immediate consequence of 14.5. We now prove 14.5. The adjoint action of T on $L(G)$ can be diagonalized. There exist elements $X_\alpha \in L(G)$ such that we have a decomposition

$$L(G) = L(T) \oplus \coprod_{\alpha \in \Sigma} K X_\alpha,$$

(see [1, Theorem, p. 317–318]).

If $\alpha, \beta \in \Sigma$ then there exist elements $N_{\alpha\beta} \in K$ such that

$$[X_\alpha X_\beta] = N_{\alpha\beta} X_{\alpha+\beta}.$$

We have $N_{\alpha\beta} = 0$ if $\alpha + \beta \notin \Sigma$. Moreover the assumption (iii) and (iv) imply that $N_{\alpha\beta} \neq 0$ if (and only if) $\alpha + \beta \in \Sigma$. In fact, if G is a semisimple adjoint group, this is a consequence of results of Chevalley ([9], taking into account that in this case G is isomorphic to one of the groups studied by Chevalley).

In the more general case considered here the result then also follows: first replace G by the derived group H of its identity component and then perform a central isogeny $\pi: H \to H_1$ onto an adjoint semisimple group. From the fact that $d\pi$ is bijective on the Lie algebra of the unipotent radical of a Borel subgroup one concludes, using that $X_\alpha \in L(H)$, that the $N_{\alpha\beta}$ are not affected by such an isogeny, whence our contention about the $N_{\alpha\beta}$.

Since D commutes with all $\operatorname{Ad}(t)$, it follows that D stabilizes the subspaces $L(T)$ and KX_α. Hence there exists a function $c: \Sigma \to K$ such that

(1) $$DX_\alpha = c(\alpha) X_\alpha.$$

§ 14. The Structure Group of a Simple J-structure

The property of $N_{\alpha\beta}$ just mentioned then shows that

(2) $$c(\alpha+\beta)=c(\alpha)+c(\beta),$$

if $\alpha, \beta, \alpha+\beta \in \Sigma$.

If $H \in L(T)$ then

$$[HX_\alpha] = d\alpha(H) X_\alpha.$$

Applying D to both sides of this formula and using (1) we find, using $DL(T) \subset L(T)$, that

(3) $$d\alpha(DH) = 0,$$

for $\alpha \in \Sigma$ and $H \in L(T)$. Since C is the intersection of the kernels of all $\alpha \in \Sigma$, we find from assumption (i), using 14.4 that $L(C)$ is the subspace

$$L(C) = \bigcap_{\alpha \in \Sigma} \mathrm{Ker}(d\alpha)$$

of $L(T)$. (3) shows that $DL(T) \subset L(C)$.

We next claim that, c being as in (1), we have

(4) $$c(-\alpha) = -c(\alpha).$$

First assume that α lies in a simple component Σ_1 of Σ which is not of type A_1. Then there exists $\beta \in \Sigma_1$ such that $\alpha+\beta \in \Sigma_1$ (choose an order on Σ_1 such that $\alpha > 0$ and let B be the basis of Σ_1 determined by that order, if $\alpha \in B$ then there is $\beta \in B$ such that $\alpha+\beta \in \Sigma_1$, if $\alpha \notin B$ there is $-\beta \in B$ such that $\alpha+\beta \in \Sigma_1$). (2) now shows that

$$c(\beta) = c(\alpha+\beta) + c(-\alpha) = c(\alpha) + c(-\alpha) + c(\beta),$$

whence (4).

If α lies in a simple component of type A_1 we have $p \neq 2$ by assumption (ii). Then $[X_\alpha X_{-\alpha}] \neq 0$ and $d\alpha([X_\alpha X_{-\alpha}]) \neq 0$, as is well-known (and follows by a simple computation in the Lie algebra of $\mathbb{SL}_2$). By (1) and (3) we find

$$0 = d\alpha(D[X_\alpha X_{-\alpha}]) = (c_\alpha + c_{-\alpha}) d\alpha([X_\alpha X_{-\alpha}]),$$

whence (4).

Now fix a basis $\{\alpha_1, \ldots, \alpha_l\}$ of Σ. From 14.4 we see that $L(C) \subset L(T)$ is the intersection of the kernels of the $d\alpha_i$. We conclude that there is $H \in L(T)$ such that

$$d\alpha_i(H) = c(\alpha_i).$$

From (2) and (4) it then follows that

$$d\alpha(H) = c(\alpha),$$

for all $\alpha \in \Sigma$. Subtracting from D the inner derivation defined by H we get a derivation of $L(G)$ into $L(C)$. This establishes 14.5.

14.7 Applications. We discuss a few cases where 14.5 applies.

(a) $G = \mathbb{P}\mathbb{G}\mathbb{L}_r$ ($= \mathbb{G}\mathbb{L}_r$ modulo its center). In that case condition (i) of 14.5 is satisfied. In fact we have $X = Y$ (as in the case of any adjoint semisimple group). The other conditions of 14.5 are also satisfied, except (ii) if $p = 2$, $r = 2$.

Apart from that exceptional case, it follows from 14.6 that all derivations of $L(G)$ are inner. In the excepted case the statement is no longer true, as one checks without trouble.

(b) $p \neq 2$ and G is a semisimple group of type B_l, C_l, D_l or F_4. In that case we have that X/Y has at most 2-torsion, so that (i) holds. The other conditions of 14.5 are also satisfied. Since C is finite, 14.6 shows that all derivations of $L(G)$ are inner.

(c) Let H be a connected and simply connected quasi-simple group of type E_6. It is known (and we have already encountered this in 11.17) that H has a irreducible representation ρ in a 27-dimensional vector space A, and that ρ is an isomorphism of G onto its image. Let G be the algebraic subgroup of $GL(A)$ generated by $\rho(G)$ and the nonzero scalar multiplications in A. G is reductive.

We claim that we now have, with the notations of 14.5, that X/Y has no torsion. We indicate the proof.

Let T_1 be a maximal torus of H contained in T. Let X_1 be the character group of T_1. The explicit information about H and ρ which one finds in [10, p. 19-07 and p. 20-05] shows that X_1 is generated by 7 elements $x_1, \ldots, x_6$ and s, subject to the relation $3s = x_1 + \cdots + x_6$. The subgroup Y_1 of X_1 spanned by the roots of H with respect to T_1 has index 3 in X_1, it is generated by the elements $x_i - x_j$, $x_i + x_j + x_k - s$ (i, j, k distinct) and s. $x_i + Y_1$ generates X_1/Y_1.

One then establishes that X is isomorphic to the subgroup of $X_1 \oplus \mathbb{Z}$ generated by the elements $(y, 0)$ ($y \in Y_1$), $(0, 3)$ and $(x_1, -1)$, and that Y corresponds under this isomorphism to the subgroup spanned by the $(y, 0)$. The verification of these facts is left to the reader. It then follows that X/Y has no torsion. Consequently, all conditions of 14.5 hold, and any derivation of $L(G)$ centralizing $\mathrm{Ad}(T)$ is sum of an inner one an a derivation of $L(G)$ into $L(C)$.

14.8. After these somewhat lengthy preparations we return to the J-structures. Let $\mathscr{S} = (V, j, e)$ be a simple J-structure. We denote by G its structure group and by G_1 the inner structure group (see 1.19). H is the automorphism group of S and C the 1-dimensional torus of nonzero scalar

§14. The Structure Group of a Simple J-structure

multiplications of V, which is a subtorus of G_1. $L(A)$ denotes again the Lie algebra of the algebraic group A.

$\mathfrak{g}$ and $\mathfrak{g}_1$ denote the structure algebra and the inner structure algebra of S, respectively. These were introduced in §4. We have

$$\mathfrak{g}_1 \subset L(G_1) \subset L(G) \subset \mathfrak{g},$$

as was established in 4.4. G_1 acts on $\mathfrak{g}$ and $\mathfrak{g}_1$, inducing trivial action on $\mathfrak{g}/\mathfrak{g}_1$ see 4.4(iii). We also have $[\mathfrak{g}, \mathfrak{g}_1] \subset \mathfrak{g}_1$. If $X \in \mathfrak{g}$, $Y \in \mathfrak{g}_1$ we put $D_X Y = [X, Y]$. D_X is a derivation of $\mathfrak{g}_1$.

We put $H_1 = H \cap G_1$, $\mathfrak{h} = \{X \in \mathfrak{g} \mid Xe = 0\}$, $\mathfrak{h}_1 = \mathfrak{g}_1 \cap \mathfrak{h}$, as in §4 (immediately before 4.7).
Then

$$L(H_1) \subset L(H) \subset \mathfrak{h}$$

and $\mathfrak{h}_1$ is an ideal in $\mathfrak{h}$.

14.9 Lemma. *Let $X \in \mathfrak{g}$. If T is a maximal torus of G_1 there exists an inner derivation D_Y ($Y \in \mathfrak{g}_1$) of $\mathfrak{g}_1$ such $D_X - D_Y$ commutes with all linear transformations $\mathrm{Ad}(t)$ ($t \in T$). The same statement is true if $\mathfrak{g}$, G_1, $\mathfrak{g}_1$ are replaced by $\mathfrak{h}$, H_1, $\mathfrak{h}_1$ or, if $\mathfrak{h}_1 \subset L(H)$, by $\mathfrak{h}$, H_1, $L(H)$.*

Since G_1 acts trivially on $\mathfrak{g}/\mathfrak{g}_1$, it follows from the complete reducibility of the action of T on $\mathfrak{g}$ that there exists a T-stable subspace $\mathfrak{m}$ of $\mathfrak{g}$ with $\mathfrak{g} = \mathfrak{g}_1 \oplus \mathfrak{m}$, such that T acts trivially on $\mathfrak{m}$. This implies the first statement. The others are proved similarly.

14.10 Lemma. (i) *Assume that $\mathfrak{g}_1 \cdot e$ has codimension d in V. Then $\dim L(G) - \dim \mathfrak{g}_1 = \dim L(H) - \dim \mathfrak{h}_1 + d$;*
(ii) *Assume that $L(G)e$ has codimension d in V. Then $\dim \mathfrak{g} - \dim L(G) = \dim \mathfrak{h} - \dim L(H) - d$.*

We have $\dim \mathfrak{g}_1 - \dim \mathfrak{h}_1 = \dim V - d = \dim G - \dim H - d = \dim L(G) - \dim L(H) - d$ (the second equality following from axiom (J3) and 4.6). This implies (i). (ii) is proved similarly.

14.11 Lemma. *Assume that $L(G)e = V$. Then*

$$L(H) = \{X \in L(G) \mid Xe = 0\}$$

and we have $\mathfrak{h}_1 \subset L(H)$.

Let $\mathfrak{h}' = \{X \in L(G) \mid Xe = 0\}$. Since $He = e$, it follows that $L(H) \in \mathfrak{h}'$. We have

$$\dim G - \dim H = \dim V = \dim L(G) - \dim \mathfrak{h}',$$

whence $\dim \mathfrak{h}' = \dim H = \dim L(H)$. This implies that $L(H) = \mathfrak{h}'$. That $\mathfrak{h}_1 \subset L(H)$ is now obvious.

14.12 Lemma. *Let $\mathscr{S}$ be a arbitrary J-structure (not necessarily simple). Assume that the standard symmetric bilinear form σ of $\mathscr{S}$ is nondegenerate.*
  (i) *If $a \in V$ is such that $\sigma(\mathfrak{g}_1 \cdot e, a) = 0$ then $P(a, e) = 0$;*
  (ii) *If $p \neq 2$ then $\mathfrak{g}_1 \cdot e = V$;*
  (iii) *If $p = 2$ and if $\mathscr{S}$ is simple then $\mathfrak{g}_1 \cdot e$ has codimension 1 in V.*

$\mathfrak{g}_1 \cdot e$ contains the elements $L(x, y)e$ ($x, y \in V$, y invertible), where L is as in §4, (4). Let $g \in G$ and apply this with $y = ge$. $g \mapsto g'$ denoting the standard automorphism, it follows, using §1, (10), 1.16(i) and §3, (10) that

$$\sigma(L(x, ge)e, a) = \sigma(P(ge)e, P(x, j(ge))a)$$
$$= \sigma(g(g')^{-1}e, g'P((g')^{-1}x, e)g^{-1}a) = \sigma(e, P(a, ge)x)$$
$$= \sigma(P(a, ge)e, x).$$

(i) is a consequence of the equality of the first and last terms. If $p \neq 2$ then $P(a, e) = 0$ implies $a = 0$ (see 10.30) and hence (i) implies (ii).

If $p = 2$ and if $\mathscr{S}$ is simple then $\mathscr{S}$ is also separable, see 13.7. Then 10.32 implies that, a being as in (i), we have $a \in Ke$. (iii) now follows from the nondegeneracy of σ.

We next prove a general statement about G_1.

14.13 Proposition. *Assume that $\mathscr{S}$ is strongly simple. Then the inner structure group G_1 of $\mathscr{S}$ coincides with the identity component G° of the structure group G.*

We know from 1.19 that G_1 is a closed normal subgroup of G°, hence of G. Moreover G_1 contains the torus C of scalar multiplications. From the results of §12 and 11.20 it follows that G° is the product of C and a quasi-simple group G', unless $\mathscr{S}$ is of type $\mathscr{M}_r$. Now $G_1 \cap G'$ is a nontrivial closed normal subgroup of G', which must be G' itself, unless $\mathscr{S}$ is of type $\mathscr{M}_r$.

To establish the same result if $\mathscr{S} = \mathscr{M}_r$ observe that in that case G' is isogeneous to $D = \mathbb{SL}_r \times \mathbb{SL}_r$. Then $G_1 \cap G'$ determines a nontrivial closed normal subgroup D_1 of D. From the explicit form of the quadratic map for $\mathscr{M}_r$, given in 2.2 we see that D_1 cannot be contained in one of the factors $\mathbb{SL}_r$ of D. Hence we must have $D_1 = D$ and $G_1 \cap G' = G'$ in all cases. It follows that $G_1 = G^\circ$.

14.14 Corollary. *If $\mathscr{S}$ is strongly simple then H and H_1 have the same identity component.*

Remark. 14.13 could also have been deduced in the case by case discussion which is given below. In fact, this discussion even shows that 14.13 holds for all simple J-structures. The cases not covered by 14.13 ($p = 2$ and $\mathscr{S}$ either $\mathscr{O}''_{2,r}$ or $\mathscr{S}_r$) are dealt with in 14.15 and 14.18.

We shall now discuss separately the situation for the various simple J-structures. We begin with the J-structures defined by quadratic forms.

§14. The Structure Group of a Simple J-structure

14.15. $\mathscr{S} = \mathscr{J}(V, Q, e)$.

We use the notations of §2 (see 2.15 and the sequel). We assume that $n = \dim V \geq 3$, that the quadratic form Q is nondegenerate and that, with the notation of 2.17, we have $Q_e \neq 0$. We know by 2.18 that G is the group of linear transformations $g \in GL(V)$ which leave invariant Q up to a nonzero factor.

Let $O(Q)$ be orthogonal group of Q. We put $SO(Q) = O(Q) \cap SL(V)$ if $p \neq 2$. If $p = 2$ and Q is nondefective (i.e. if the associated alternating bilinear form is nondegenerate), we denote by $SO(Q)$ the rotation subgroup of $O(Q)$ (see [12, p. 65]) and if $p = 2$ and Q is defective we write $SO(Q) = O(Q)$. We have $G = O(Q)C$. One knows that $SO(Q)$ is a connected closed normal subgroup of $O(Q)$, of index ≤ 2. $\mathscr{S}$ is now isomorphic to one of the J-structures $\mathscr{O}_{2,r}(\text{char}(K) \neq 2, r \geq 5)$, $\mathscr{O}'_{2,r}(\text{char}(K) = 2, r \geq 3)$, $\mathscr{O}''_{2,r}(\text{char}(K) = 2, r \geq 2)$.

It follows from 13.5 and 14.13 that $G_1 = SO(Q)C$ unless perhaps if $\mathscr{S}$ is isomorphic to $\mathscr{O}''_{2,r}$. But in that case we have, as in the proof of 14.13, that G is the product of C and a quasi-simple group (see [10, p. 22-04 and 22-05]), so that the argument of 14.13 can again be applied. It follows that $G_1 = SO(Q)C$ in all cases.

If $p \neq 2$ then $SO(Q)$ is the subgroup of $O(Q)$ whose elements have determinant 1. Since $O(Q) \cap C = \{\pm \text{id}\}$ we conclude that we have $G = G^\circ$ if $n = \dim V$ is odd and $(G : G^\circ) = 2$ if n is even.

If $p = 2$ then $O(Q) \cap C = \{\text{id}\}$, whence $(G : G^\circ) = (O(Q) : SO(Q))$. If Q is nondefective this index is 2 and if Q is defective the index is 1 (for then $O(Q)$ is topologically isomorphic to a symplectic group, see [loc. cit., p. 22-05]). Since Q is defective if and only if n is odd we obtain the same result as for $p \neq 2$.

By 4.6, H is the subgroup of G whose elements fix e. Let $p \neq 2$. Then H is (topologically) isomorphic to $O(Q_1)$, where Q_1 is the restriction of Q to the orthogonal complement of Ke. Hence H is not connected and $(H : H^\circ) = 2$.

If $p = 2$ and if Q is nondefective we find again that H is isomorphic to $O(Q_1)$, where Q_1 is a nondegenerate quadratic form in one dimension less. Since now $O(Q_1)$ is isomorphic to a symplectic group, it follows that H is connected.

If $p = 2$ and Q is defective, one checks that H is isomorphic to the subgroups of a symplectic group which fixes a nonzero vector, in its natural representation. Such a group is again connected.

It also follows that H° is semisimple if $n > 3$, unless $p = 2$ and Q is defective (the verification of the last statement is left to the reader). Observe that $\dim G = \frac{1}{2} n(n-1) + 1$, $\dim H = \dim G - n = \frac{1}{2}(n-1)(n-2)$.

We next come to the discussion of the various Lie algebras in this case. The definition of $\mathfrak{g}$ in §4 shows that it is the space of all linear

transformations X of V such that there exists a linear transformation Y with
$$X\bar{x} - Yx = Q(x)^{-1}(Q(x, \overline{X\bar{x}})x) \quad (Q(x) \neq 0),$$
the notations being as in §2.

The right-hand side of the equality must clearly be a polynomial map of V into itself. Let F be the rational function on V defined by
$$F(x) = Q(x)^{-1} Q(x, \overline{X\bar{x}}) \quad (Q(x) \neq 0).$$
If $(e_i)_{1 \leq i \leq n}$ is a basis of V then
$$(x_1, \ldots, x_n) \mapsto \sum_{i=1}^{n} (F(\Sigma_j x_j e_j) x_i e_i)$$
defines a polynomial map of K^n into V.

It follows that the denominator of F divides all coordinate functions $(x_1, \ldots, x_n) \mapsto x_i$, from which one concludes that F is a polynomial function, which then must be a constant. Consequently, $\mathfrak{g}$ is the space of all linear transformations X of V such that $Q(x, \overline{X\bar{x}})$ is a multiple of $Q(x)$, which coincides with the space of all $X \in \text{End}(V)$ such that there exists $a \in K$ with

(5) $$Q(x, Xx) = aQ(x).$$

First assume that the symmetric bilinear form $Q(\,,\,)$ is nondegenerate. Let $X \in \text{End}(V)$ satisfy (5). Define matrices $S = (s_{ij})$ and $T = (t_{ij})$ by
$$s_{ij} = Q(e_i, e_j), \quad t_{ij} = Q(Xe_i, e_j) \quad (1 \leq i, j \leq n).$$
X is uniquely determined by T. It follows that (5) is equivalent to the following properties of T

(6) $$\begin{cases} t_{ii} = aQ(e_i) \\ T + {}^tT = S. \end{cases}$$

Let $A = (a_{ij})$ be a fixed matrix with $a_{ii} = Q(e_i)$, $A + {}^tA = S$. It follows from (6) that

(7) $$T = aA + U,$$

where U is an alternating matrix. Conversely, if T has the form (7) then (6) holds. It follows that
$$\dim \mathfrak{g} = \tfrac{1}{2} n(n-1) + 1 = \dim G.$$
Since $L(G) \subset \mathfrak{g}$ this shows that $\mathfrak{g} = L(G)$.

From §2, (12) we see that the nondegeneracy of $Q(\,,\,)$ implies that the standard symmetric bilinear form σ is nondegenerate. 14.12(ii) then shows that $\mathfrak{g}_1 \cdot e = V$ if $p \neq 2$, whence $L(G)e = V$ in that case.

§14. The Structure Group of a Simple J-structure   145

We claim that $L(G)e = V$ also if $p = 2$. Let $e = \sum_{i=1}^{n} a_i e_i$ and assume that $f = \sum_{i=1}^{n} b_i e_i$ is such that $Q(Xe, f) = 0$ for all $X \in L(G)$. This means that

$$\sum_{i,j} a_i b_j t_{ij} = 0$$

for all matrices T of the form (7). One easily concludes that all b_j have to be 0, whence $f = 0$. Consequently $L(G)e = V$. Now 14.10(ii) implies that we also have $\mathfrak{h} = L(H)$.

We next investigate $\mathfrak{g}_1$. From §4, (4) and §2, (13) we see that

$$L(x, y) z = P(x, z) y = Q(x, \bar{y}) z + Q(z, \bar{y}) x - Q(x, z) \bar{y}.$$

First let $p \neq 2$. Choose now (e_i) to be an orthogonal basis of V for Q (i.e. such that $Q(e_i e_j) = 0$ if $i \neq j$), with $e_i = \pm e_i$. This is possible.

Then $L(e_i, e_i)$ is a nonzero scalar multiple of the identity and the $L(e_i, e_j)$ with $i < j$ are linearly independent elements of $\mathrm{End}(V)$. It follows that

$$\dim \mathfrak{g}_1 = \tfrac{1}{2} n(n-1) + 1 = \dim L(G),$$

whence $\mathfrak{g}_1 = L(G)(= \mathfrak{g})$. 14.10(i) and 14.11 then imply that $\mathfrak{h}_1 = L(H)$.

If $p = 2$ then $n = 2r$ is even. Let now $(e_i)_{1 \leq i \leq 2r}$ be a symplectic basis of V for Q, i.e. such that

$$(8) \qquad Q(e_i, e_j) = \begin{cases} 0 & \text{if } |i-j| \neq r \\ 1 & \text{if } |i-j| = r, \end{cases}$$

with $e = e_1 + e_r$ (such a basis exists). A simple computation then shows that the $L(e_i, e_j)$ span a subspace of $\mathrm{End}(V)$ of dimension $\tfrac{1}{2} n(n-1) = \dim L(G) - 1$. Hence $\mathfrak{g}_1$ has codimension 1 in $\mathfrak{g}$. Since $\mathfrak{g}_1 \cdot e$ has codimension 1 in V by 14.12(iii) we conclude from 14.10(i) and 14.11 that $\mathfrak{h}_1 = L(H)$.

There remains the case that $Q(\,,\,)$ is degenerate, i.e. that $p = 2$ and Q is defective. In that case n is odd. Let f span the space of vectors orthogonal to all of V. Since Q is nondegenerate we have $Q(f) \neq 0$. Also $e \notin Kf$.

Let $X \in \mathrm{End}(V)$ satisfy (5). From (5) we find

$$Q(Xf, x) = a Q(f, x) - Q(f, Xx) = 0,$$
$$a Q(f) = Q(Xf, f) = 0.$$

Hence Xf is a multiple of f and $a = 0$.

Let l be a linear function on V. Then the linear transformation X_l of V defined by

$$(9) \qquad X_l x = l(x) f$$

is in $\mathfrak{g}$. Considering the nondegenerate alternating form induced by $Q(\,,\,)$ on $V/Kf \times V/Kf$ one finds from the preceding remarks that

$$\dim \mathfrak{g} = \tfrac{1}{2}(n-1)(n-2) + n = \dim G,$$

so that we have again $\mathfrak{g} = L(G)$.

(5) implies that $Q(Xe, e) = 0$ for all $X \in \mathfrak{g}$. Since $e \notin Kf$ this shows that $L(G)e \neq V$. 14.10(ii) now implies that $\mathfrak{h} \neq L(H)$. Hence H is not smooth (in the sense of 4.9).

We finally discuss $\mathfrak{g}_1$ and $\mathfrak{h}_1$ in this case. Let $(e_i)_{0 \leq i \leq 2r}$ be a basis of V such that $e_0 = f$, that $(e_i)_{1 \leq i \leq 2r}$ satisfies (8) and that $e = e_1 + e_r$ (such a basis exists). Then the $L(e_i, e_j)$ span a space of dimension $\tfrac{1}{2}n(n-1) = \dim L(G) - 1$. So $\mathfrak{g}_1$ has codimension 1 in $L(G)$. Let X_l be as in (9), with $l(f) \neq 0$. One proves that this X_l cannot lie in $\mathfrak{g}_1$ (the proof being left to the reader), whence $L(G) = \mathfrak{g}_1 + KX_l$. Since $f \in \mathfrak{g}_1 \cdot e$, this shows that $\mathfrak{g}_1 \cdot e = L(G)e$. 14.10(i) and 14.11 now imply that $\dim L(H) = \dim \mathfrak{h}_1$ and that $L(G)e$ has codimension 1 in V. By 14.10(ii) we have that $L(H)$ has codimension 1 in $\mathfrak{h}$.

This concludes the discussion of the J-structures defined by quadratic forms.

14.16. $\mathscr{S} = \mathscr{M}_r$ $(r \geq 1)$.

$V = \mathbb{M}_r$, the space of $r \times r$ matrices. If $X, Y \in \mathbb{GL}_r$, let $\phi(X, Y)$ be the linear transformation of V defined by

$$\phi(X, Y)Z = XZY^{-1}.$$

It follows from 11.20 (see the proof of 12.7) that G° is the group of all transformations $\phi(X, Y)$. Since $\mathscr{M}_r$ is strongly simple it follows from 14.13 that $G_1 = G^\circ$.

Consider the homomorphism ϕ (of algebraic groups) of $\mathbb{GL}_r \times \mathbb{GL}_r$ into $GL(V)$. We have $\phi(X, Y) = \mathrm{id}$ if and only if $X = Y = x \cdot \mathrm{id}$ $(x \in K^*)$.

The differential $d\phi$ is the linear map $\mathfrak{gl}_r \times \mathfrak{gl}_r \to \mathrm{End}(\mathbb{M}_r)$ defined by

(10) $$d\phi(X, Y)(Z) = XZ - ZY.$$

Clearly $d\phi(X, Y) = 0$ if and only if $X = Y = x \cdot \mathrm{id}$ $(x \in K)$, whence $\dim \mathrm{Ker}\,\phi = \dim \mathrm{Ker}\,d\phi$. Consequently ϕ is a separable homomorphism of algebraic groups (see [1, 17.3, p. 75]), and ϕ induces an isomorphism of algebraic groups of $\mathbb{GL}_r \times \mathbb{GL}_r/A$ onto G°, where A is the subtorus consisting of all $(x \cdot \mathrm{id}, x \cdot \mathrm{id})$ $(x \in K^*)$.

By 4.6 it follows that $H \cap G^\circ$ is the group of all $\phi(X, X)$ $(X \in \mathbb{GL}_r)$. An argument similar to the one just given shows that $H \cap G^\circ$ is isomorphic to $\mathbb{PGL}_r$ ($= \mathbb{GL}_r$ modulo its center) and is connected.

Let $h \in H - H \cap G^\circ$. The inner automorphism $\mathrm{Int}_H(h)$ of H defined by h then induces an outer automorphism of $H \cap G^\circ$. Since the group of

§14. The Structure Group of a Simple J-structure

inner automorphisms of $\mathbb{PGL}_r$ has index 2 in the group of all automorphisms (in the sense of algebraic groups), as follows e.g. from [10, Cor. 3, p. 24-04], we see that $(H:H^\circ) \leq 2$.

One furthermore sees, using the explicit description of G° which we gave, that every coset of G/G° is represented by an element of H/H°. Since the transformation $\varepsilon: X \mapsto {}^t X$ is clearly in H, we have $(G:G^\circ) = (H:H^\circ) = 2$.

Consequently, G is the group of linear transformations generated by the $\phi(X, Y)$ and ε, and H is the subgroup generated by the $\phi(X, X)$ and ε. Also $H_1 = H^\circ$.

We next discuss the Lie algebras. From the explicit description of G° we see that $L(G)$ is the Lie algebra of all linear transformations $d\phi(X, Y)$ of the form (10). It is then clear that $L(G) e = V$. From 2.2 and §4, (4) we conclude that

$$L(X, Y) Z = (XY) Z + Z(YX) \quad (X, Y, Z \in \mathbb{M}_r).$$

Let (e_{ij}) be the canonical basis of $\mathbb{M}_r$. Then

$$L(e_{ik}, e_{kj}) = d\phi(e_{ij}, 0), \quad L(e_{kj}, e_{ik}) = d\phi(0, -e_{ij}) \quad (i \neq j),$$
$$L(e_{ij}, e_{ji}) = d\phi(e_{ii}, e_{jj})$$
$$L(X, e) = d\phi(X, -X).$$

It follows from these relations that $\mathfrak{g}_1$ has codimension ≤ 1 in $L(G)$ and that $L(G) = \mathfrak{g}_1 + K \cdot \mathrm{id}$ if p does not divide r. In the latter case we have, if moreover $p \neq 2$, that $L(\mathrm{id}, \mathrm{id}) = 2 \cdot \mathrm{id}$, whence $L(G) = \mathfrak{g}_1$. It also follows that $\mathfrak{g}_1 \cdot e = V$ if $p \neq 2$. If $p = 2$ then 14.12(iii) shows that $\mathfrak{g}_1 \cdot e$ has codimension 1 in V (by 2.4 the standard symmetric bilinear form is nondegenerate, so that the hypothesis of 14.12 is satisfied).

If $p \neq 2$ does not divide r, then we find from 14.10(i) and 14.11 that $\mathfrak{h}_1 = L(H) \simeq L(\mathbb{PGL}_r)$. If $p = 2$ the same follows. Application of 14.9 and 14.7(a) shows that if $p \neq 2$ or $r \neq 2$, $\mathfrak{h}$ is spanned by $L(H)$ and linear transformations $D \in \mathfrak{h}$ which can be assumed to commute with all $d\phi(X, X)$, where X is a diagonal matrix. If $p \neq 2$ divides r, then the last statement of 14.9 shows, using 14.11, that the same holds.

It follows that there exist $a_{ij} \in K$ with

$$D e_{ij} = a_{ij} e_{ij} \quad (1 \leq i, j \leq r).$$

From the fact that D commutes with all $d\phi(e_{kl}, e_{kl})$ one concludes that all a_{ij} are equal, hence D is a scalar multiple of the identity. Since $De = 0$, we have $D = 0$.

It follows that $\mathfrak{h} = L(H)$ if $p \neq 2$ or $r \neq 2$. In the exceptional case: $\mathcal{M}_2$ in characteristic 2 we have that $\mathcal{M}_2 \simeq \mathcal{O}'_{2,2}$ (see the proof of 12.7), and by the discussion of 14.15 it remains true that $\mathfrak{h} = L(H)$. Hence this holds in all cases. 14.10(ii) then shows that $\mathfrak{g} = L(G)$. If $p \neq 2$ divides r, one may check that $L(H) \neq \mathfrak{h}_1$, $L(G) \neq \mathfrak{g}_1$.

14.17. $\mathscr{S} = \mathscr{S}_r$ $(p \neq 2, r > 2)$.
$V = \mathbb{S}_r$, the space of symmetric $r \times r$ matrices. From 11.20 we infer that G° is the group of linear transformations $\phi(X)$ $(X \in \mathbb{GL}_r)$ of V, with

$$\phi(X) Y = X Y \cdot {}^t X \qquad (Y \in V).$$

ϕ defines a homomorphism of algebraic groups $\mathbb{GL}_r \to G^\circ$. Its differential is the homomorphism $\mathfrak{gl}_r \to L(G)$ defined by

(11) $$d\phi(X)(Y) = XY + Y \cdot {}^t X.$$

Since $\mathscr{S}_r$ is strongly simple, 14.13 shows that $G_1 = G^\circ$.

Since $p \neq 2$, it follows that $d\phi$ is bijective. Hence ϕ is separable and we conclude that ϕ induces an isomorphism of algebraic groups of $\mathbb{GL}_r/\{\pm \mathrm{id}\}$ onto G°.

In a similar manner one sees that there is a separable surjective homomorphism ψ of G°/C onto $\mathbb{PGL}_r$.

Let $g \in G - G^\circ$. The inner automorphism $\mathrm{Int}_G(g)$ induces an automorphism of the algebraic group G°, and then $\psi \circ \mathrm{Int}_G(g)$ defines an automorphism of $\mathbb{PGL}_r$. It follows, using the fact that the coset of outer automorphisms of the algebraic group $\mathbb{PGL}_r$ modulo inner automorphisms is represented by the one coming from the outer automorphism $X \mapsto {}^t X^{-1}$ of $\mathbb{GL}_r$ (which we also used in 14.16), that we may assume that there exists a character χ of $\mathbb{GL}_r$ such that

$$g(X \cdot {}^t X) = g(\phi(X) e) = \chi(X) \phi({}^t X^{-1}) g(e),$$

or, putting $A = g e$,

(12) $$g(X \cdot {}^t X) = \chi(X) {}^t X^{-1} A X^{-1},$$

for $X \in \mathbb{GL}_r$.

Take for X the diagonal matrix $\mathrm{diag}(x_1, \ldots, x_n)$ $(x_i \in K^*)$. Then there exists an integer d such that $\chi(X) = (x_1 \ldots x_n)^d$ and one sees that (12) is only possible if $r = 2$, a case which we have excluded. Consequently $G = G^\circ$.

From 4.6 we see that H is isomorphic to the orthogonal group $\mathbb{O}_r = \{X \in \mathbb{GL}_r | X \cdot {}^t X = 1\}$. Hence H° has index 2 in H. We have $H_1 = H$.

Next the Lie algebras. $L(G)$ is the Lie algebra of the $d\phi(X)$ given by (11). It is obvious that $L(G) e = V$. By §4, (4) we have, using what was established in §2 that now

(13) $$L(X, Y) Z = (XY) Z + Z \cdot {}^t(XY).$$

§ 14. The Structure Group of a Simple J-structure

Since, as one easily checks, every matrix is a sum of products of two symmetric ones, it follows from (13) that $\mathfrak{g}_1$ contains all linear transformations of the form (11), whence $\mathfrak{g}_1 = L(G)$. By 14.10(i) and 14.11 we see that $\mathfrak{h}_1 = L(H)$.

Now the algebraic group $H°$ is isomorphic to a semisimple group of type B_l or C_l. From 14.9 and 14.7(b) we see that $\mathfrak{h}$ is again spanned by $L(H)$ and linear transformations $D \in \mathfrak{h}$ which centralize all $\phi(X)$, where X is diagonal. An argument like the one given at the end of 14.16 now shows that $\mathfrak{h} = L(H)$. From 14.10(ii) we obtain that $\mathfrak{g} = L(G)$.

14.18. $\mathscr{S} = \mathscr{S}_r$ $(p=2, r>2)$.

We use the notations introduced at the beginning of 14.17. We can no longer apply 11.20 to identify $G°$. At any rate, we still have the homomorphism $\phi: \mathbb{GL}_r \to G°$. $d\phi$ is no longer injective. One easily checks that Ker $d\phi$ consists of the scalar multiples of the identity.

As in 14.17 one sees that $\mathfrak{g}_1$ consists of the linear transformations of the form $d\phi(X)$. From what we observed about Ker $d\phi$ it follows that $\mathfrak{g}_1$ is isomorphic to $L(\mathbb{PGL}_r)$. Hence dim $\mathfrak{g}_1 = r^2 - 1$. One establishes that the identity map of V is not of the form (11).

We have $\mathfrak{g}_1 + K \cdot \mathrm{id} \subset L(G) \subset \mathfrak{g}$. We shall show that $\mathfrak{g} = \mathfrak{g}_1 + K \cdot \mathrm{id}$. It will then follow that $\mathfrak{g} = L(G)$, dim $G = r^2$, which will imply that $G = \phi(\mathbb{GL}_r)$.

So let $D \in \mathfrak{g}$. Using 14.9 and 14.7(a) we see that we may assume D commutes with all $d\phi(X)$. It then follows that $((e_{ij})$ being the canonical basis of $\mathbb{M}_r$) D stabilizes all subspaces $K(e_{ij}+e_{ij})$ $(i \neq j)$ of $\mathbb{S}_r$ and also the subspace of diagonal matrices. In particular, D stabilizes the space $\mathbb{A}_r$ of alternating $r \times r$ matrices and commutes with the action of $\mathfrak{g}_1$ on $\mathbb{A}_r$.

It follows that the restriction of D to $\mathbb{A}_r$ is a scalar multiple of the identity, which we may assume to be 0. Hence DX is diagonal for all X. By 4.5 we have

(14) $$DX^2 = X(DX) + (DX)X + XD(e)X \quad (X \in \mathbb{S}_r),$$

from which one concludes, by taking $X \in \mathbb{A}_r$, that $D(e)$ is a diagonal matrix such that $XD(e)X$ is diagonal for all $X \in \mathbb{A}_r$, which can only be if $D(e) = 0$.

(14) then implies that $DX = 0$ for all diagonal X. Hence $D = 0$. This establishes our claim that $\mathfrak{g} = \mathfrak{g}_1 + K \cdot \mathrm{id}$. As we saw, this implies that $G°$ is the group of $\phi(X)$ $(X \in \mathbb{GL}_r)$. The homomorphism ψ of $G°/C$ onto $\mathbb{PGL}_r$ of 14.17 is an isomorphism also in this case, and can again be used to establish $G = G°$.

$G°$ is now the product of its center and a quasi-simple group, from which one concludes as in the first paragraph of the proof of 14.13, that $G_1 = G°$.

From what we have established so far, it follows that $\mathfrak{g}_1$ has codimension 1 in $L(G)$. Since $\mathfrak{g}_1 \cdot e$ is the space $\mathbb{A}_r$ of alternating $r \times r$ matrices, we conclude that $L(G) e = \mathbb{A}_r + K e$. Hence $L(G) e$ has codimension $r-1$ in V and $\mathfrak{g}_1 \cdot e$ has codimension r. From 14.10(ii) we see that $\dim \mathfrak{h} = \dim L(H) + r - 1$. Consequently $\mathfrak{h} \neq L(H)$, so that H is not smooth. 14.10(i) shows that $\dim \mathfrak{h}_1 = \dim L(H) + r - 1$, whence $\mathfrak{h} = \mathfrak{h}_1$.

Observe, finally, that in this case H is the subgroup of $\mathbb{GL}_r$ which leaves invariant a nondegenerate, nonalternating symmetric bilinear form on $K^r \times K^r$. Such a group is a non-reductive linear algebraic group (as follows from [12, p. 20]).

14.19. $\mathscr{S} = \mathscr{A}_r$ $(r > 2)$.

The discussion in that case is quite similar to that of the preceding cases. One first proves that $G = G^\circ$, as in the case $\mathscr{S}_r$. Moreover, H is isomorphic to the projective symplectic group $\mathbb{PSp}_{2r}$.

If $p \neq 2$ then $L(G) = \mathfrak{g}_1$, if $p = 2$ then $L(G) \neq \mathfrak{g}_1$ and $L(G) = \mathfrak{g}_1 + K \cdot \mathrm{id}$. We have $L(G) e = V$ in all cases. If $p \neq 2$ then one finds, as in the case of $\mathscr{S}_r$, that $\mathfrak{h} = L(H)$ and $\mathfrak{g} = L(G)$. If $p = 2$ then, using $\mathfrak{g}_1 \simeq L(\mathbb{PGL}_{2r})$, as in 14.18, one proves that $\mathfrak{g} = L(G)$, whence $\mathfrak{h} = L(H)$.

14.20. $\mathscr{S} = \mathscr{E}_3$.

We first recall some facts from Section 5. We have now $V = (\mathbb{M}_3)^3$.

If $x = (x_0, x_1, x_2)$, $y = (y_0, y_1, y_2) \in V$ then the standard symmetric bilinear form is given by

$$\sigma(x, y) = \tau(x_0 y_0 + x_1 y_1 + x_2 y_2),$$

whence τ is the trace in $\mathbb{M}_3$. σ is nondegenerate. 14.13 implies that $G_1 = G^\circ$.

The standard automorphism $\alpha: g \mapsto g'$ of G is given by

(15) $$\sigma(g x, g' x) = \sigma(x, y).$$

From 11.20 it follows that G° is the product of a simply connected quasi-simple algebraic group G' of type E_6 and the center C. Since G' is the commutator subgroup of G°, it follows that α stabilizes G'.

Assume that $\alpha g = h g h^{-1}$ for some $h \in G'$, for all $g \in G'$. Put $f(x, y) = \sigma(x, h y)$. Then it follows from (15) that $f(g x, g y) = f(x, y)$ $(g \in G')$, which would imply that the representation of G' in V was equivalent to its contragredient. But the 27-dimensional irreducible representation of G' in V is not equivalent to its contragredient (as follows from the results established in [10, p. 20-10, p. 20-05]), hence we arrive at a contradiction. So α is an outer automorphism of G'.

One knows that the group of inner automorphism of G' has index 2 in the group of all its automorphisms (in the sense of algebraic groups),

§ 14. The Structure Group of a Simple J-structure

as follows from [10, Cor. 3, p. 24-04]. It follows that α represents the coset of outer automorphisms. Let $h \in G - G^\circ$, let β be the automorphism of G' induced by $\text{Int}_G(h)$. We may then assume that $\beta = \alpha$. That $G = G^\circ$ now follows by an argument similar to the one used in 14.17 to prove the corresponding assertion for $\mathcal{S}_r$.

By 12.3(iii) we know that G is the group of linear transformations which leave invariant the norm N of $\mathcal{S}$, which is a cubic form. It follows that the center $G' = G' \cap C$ of G' is a cyclic group of order 3 if $p \neq 3$ and is reduced to the identity if $p = 3$. Moreover $G' = G \cap SL(V)$.

We next study H. Let

$$G'_\alpha = \{ g \in G' | \alpha g = g \}.$$

From $G = G' \cdot C$ one concludes that G'_α is also the group of all elements of G which are fixed by α. Since $\mathcal{S}$ is separable if $\text{char}(K) = 2$ (see 7.9), we find from 7.14 that H is of finite index in G'_α.

We shall now establish that G'_α is a connected group, which will imply that $H = G'_\alpha$. In order to prove this fact, we shall invoke some results about automorphisms of semisimple group due to Steinberg. To be in a position to apply these results we have to discuss the root system of G'.

14.21 The root system of G'. We keep the previous notations. Let S be the 2-dimensional diagonal torus in $\mathbb{SL}_3$. Define a rational homomorphism ϕ of S^3 into $GL(V)$ by

(16) $\qquad \phi(t_0, t_1, t_2)(x_0, x_1, x_2) = (t_0 x_0 t_1^{-1}, t_1 x_1 t_2^{-1}, t_2 x_2 t_0^{-1}).$

From 5.12 we see that $\phi(t_0, t_1, t_2) \in G$, and since clearly $\phi(t_0, t_1, t_2) \in SL(V)$, this is an element of $G' = G \cap SL(V)$.

Put $T = \phi(S^3)$. This is a 6-dimensional torus in G', which must be a maximal one, since G' has rank 6. Observe that

(17) $\qquad \alpha(\phi(t_0, t_1, t_2)) = \phi(t_1, t_0, t_2),$

as follows from what was established in 5.12.

We define characters x_{ij} ($0 \leq i, j \leq 2$) of S^3 as follows. Let $t_i = \text{diag}(t_{0i}, t_{1i}, t_{2i})$, then

$$(t_0, t_1, t_2)^{x_{ij}} = t_{ij}.$$

Let X be the character group of T, which we identify with a subgroup of the character group of S^3 (via the transposed of ϕ).

It then follows from (16) that the weights of T in V are the characters

$$x_{ij} - x_{\sigma(i), k}$$

of T, where $\sigma(i) \equiv i+1 \mod 3$. The corresponding weight space has dimension 1, let x_{ijk} be a nonzero vector in that space.

Put $\rho(0)=1$, $\rho(1)=0$, $\rho(2)=2$. By 4.4 we have

$$L(x_{ijk}, x_{qrs}) \in L(G)$$

and, using 1.16(i), one finds that

$$t \circ L(x_{ijk}, x_{qrs}) \circ t^{-1} = t^a L(x_{ijk}, x_{qrs}) \quad (t \in T),$$

where

$$a = x_{ij} - x_{\sigma(i),k} + x_{\rho(q),r} - x_{\rho(q),s}.$$

We shall now exhibit nonzero elements $L(x_{ijk}, x_{qrs})$. Define linear maps $\phi_i: \mathbb{M}_3 \to V$ by $\phi_0 a = (a, 0, 0)$, $\phi_1 a = (0, a, 0)$, $\phi_2 a = (0, 0, a)$.

14.22 Lemma. *Let $a, b \in \mathbb{M}_3$.*
(i) *If $ab \neq 0$ or $ba \neq 0$ then $L(\phi_1 a, \phi_2 b) \neq 0$, $L(\phi_2 a, \phi_1 b) \neq 0$;*
(ii) *If there exists $x \in \mathbb{M}_3$ with $b \times a \neq \tau(b \, x) a$ then $L(\phi_0 a, \phi_1 b) \neq 0$;*
(iii) *If there exists $x \in \mathbb{M}_3$ with $a \times b \neq \tau(b \, x) a$ then $L(\phi_2 a, \phi_0 b) \neq 0$.*

Using the notations of §5, it follows from §5, (7) and the definitions of $L(x, y) z$ (see §4, (4)) that

$$L(x, y) z = \sigma(x, y) z + \sigma(y, z) x - y \times (x \times z).$$

From §5, (15) we find, if $x = (x_0, x_1, x_2) \in V$,

$$L(\phi_1 a, \phi_2 b) x = -(b \, a \, x_0, x_1 \, a \, b, y),$$

for some $y \in \mathbb{M}_3$. This implies (i). The proofs of (ii) and (iii) are similar.

14.22 shows that we have $L(x_{ijk}, x_{qrs}) \neq 0$ in the following cases:

$$i=0, \ q=1, \ j \neq q,$$
$$i=2, \ q=0, \ k \neq r,$$
$$i=1, \ q=2, \ k=r \text{ or } j=s,$$
$$i=2, \ q=1, \ k=r \text{ or } j=s.$$

As a consequence one finds that the following characters of T are roots
$$\alpha_{ijk} = x_{0i} + x_{1j} + x_{2k} \quad (0 \leq i, j, k \leq 2),$$
$$-\alpha_{ijk},$$
$$\beta_{ijk} = x_{ij} - x_{ik} \quad (0 \leq i, j, k \leq 2, j \neq k).$$

The number of these being 72, which is the number of roots of a root system of type E_6, we conclude that $\{\alpha_{ijk}, -\alpha_{ijk}, \beta_{ijk}\}$ is the root system Σ of G' with respect to T. From (17) we see that the automorphism α

§ 14. The Structure Group of a Simple J-structure

stabilizes T. Let α^* denote the transposed map of the character group X of T. Then (17) implies that

(18) $\quad \begin{cases} \alpha^*(\alpha_{ijk}) = \alpha_{jik}, \\ \alpha^*(\beta_{0jk}) = \beta_{1jk}, \quad \alpha^*(\beta_{0jk}) = \beta_{1jk}, \quad \alpha^*(\beta_{2ij}) = \beta_{2jk}. \end{cases}$

14.23 Lemma. *α fixes T and a Borel subgroup B of G', containing T.*

We have already seen that α fixes T. Put $\alpha_1 = \alpha_{100}$, $\alpha_2 = \alpha_{110}$, $\alpha_3 = \alpha_{212}$, $\alpha_4 = \alpha_{001}$, $\alpha_5 = \alpha_{122}$, $\alpha_6 = \alpha_{010}$. Then $\alpha^*(\alpha_1) = \alpha_6$, $\alpha^*(\alpha_2) = \alpha_2$, $\alpha^*(\alpha_3) = \alpha_5$, $\alpha^*(\alpha_4) = \alpha_4$ (and $(\alpha^*)^2 = \mathrm{id}$). One checks that $\alpha_1 + \alpha_3$, $\alpha_2 + \alpha_3$, $\alpha_3 + \alpha_4$, $\alpha_4 + \alpha_5$, $\alpha_5 + \alpha_6$ are again in Σ and no other sums $\alpha_i + \alpha_j$ ($i \neq j$).

This implies that the Coxeter graph of the subsystem $\Delta = \{\alpha_1, \ldots, \alpha_6\}$ of Σ, as defined in 14.1, is the Dynkin graph of a root system of type E_6 (notice that the signs $\varepsilon_{\alpha_i \alpha_j}$ of 14.1 are now all 1, for with the notations of 14.1, $\alpha_i + \alpha_j \in \Sigma$ implies that $(\alpha_i | \alpha_j) < 0$). By 14.2 it follows that Δ is a basis of Σ.

Since α^* stabilizes Δ, the corresponding set Σ^+ of positive roots is also stabilized by α^*, which implies that α fixes the Borel subgroup $B \supset T$ defined by Σ^+. This proves 14.23.

Remark. The use of 14.2 can be avoided by establishing explicitly that Δ is a basis of Σ.

14.24 Determination of H. By a theorem of Steinberg [30, Th. 8.2, p. 52] one knows that α is an automorphism of a connected, semisimple, simply connected linear algebraic group G' such that α fixes a Borel subgroup B of G' and a maximal torus $T \subset B$, then the group of fixed points G'_α is a connected reductive group. By 14.23 we are in these circumstances here, consequently G'_α is connected and, as we saw before, this shows that $H = G'_\alpha$ is connected. Observe that $\dim H = \dim G - \dim V = 79 - 27 = 52$.

From [30, Remark 8.3(a), p. 56] we see that a maximal torus of H is

$$T_\alpha = \{t \in T \mid \alpha t = t\}.$$

Using the explicit form of α^*, given by (18), one finds that $\dim T_\alpha = 4$. Hence rank $H = 4$.

Let $R = X \otimes_{\mathbb{Z}} \mathbb{R}$, then $\Sigma \subset R$, $\Delta \subset R$. α^* induces an automorphism of R, also to be denoted by α^*, fixing Δ and Σ. Equip $R \times R$ with a positive definite scalar product, invariant under the Weyl group. Let V' be the set of fixed points of α^* in V.

There exists a root system Σ' in V' such that the root system of H with respect to T_α is isomorphic to a subsystem of Σ'. See [30, Remark 8.3(a), p. 56], Σ' is defined in [loc.cit., 1.32, p. 15].

The rule for finding a basis of Σ', given in [loc.cit., 1.30, p. 14] shows that Σ' is a simple system of type F_4, hence Σ' has 48 elements. Consequently the root system of H with respect to T_α has at most 48 elements. But since $\dim H = 52$, it follows that Σ' equals the latter root system. Hence H is a connected semisimple group of type F_4.

14.25. We finally discuss the Lie algebras in the case of $\mathscr{E}_3$. We first establish that $L(G)e = V$. If $p \neq 2$ then since σ is nondegenerate, 14.12(ii) shows that $\mathfrak{g}_1 \cdot e = V$, whence $L(G)e = V$.

If $p = 2$ then $\mathfrak{g}_1 \cdot e$ has codimension 1 by 14.12(iii). It also follows from the proof of 14.12, using §3, (10), that $\sigma(\mathfrak{g}_1 \cdot e, e) = 0$. As $\sigma(e,e) = 3$, this shows that $e \notin \mathfrak{g}_1 \cdot e$. But since $e \in L(G)e$ (for the identity is in $L(G)$), we again have $L(G)e = V$.

14.26 Lemma. (i) $L(G)/L(C)$ *is a simple Lie algebra if* $p \neq 3$;
(ii) $L(H)$ *is a simple algebra if* $p \neq 2$.

The derived group G' of G is a quasi-simple group of type E_6 and $G = G' \cdot C$. If $p \neq 3$ we have that $G' \cap C$ has order 3. Consequently the canonical morphism $G' \to G/C$ is then separable, whence $L(G)/L(C) \simeq L(G')$. The latter is one of the Lie algebras introduced by Chevalley in [9].

It is known that in the present situation ($p \neq 3$, type E_6) such an algebra is simple. This is proved in [29, 2.6(5), p. 1120], the assumption $p \neq 3$ coming in to establish that $L(G')$ has no center. $L(H)$ is a Lie algebra of Chevalley's kind of type F_4. The same reference gives that (ii) holds.

We can now determine $\mathfrak{g}_1$. It is an ideal in $L(G)$, which contains $L(C)$ if $p \neq 2$ (since $L(e,e) = 2 \cdot \mathrm{id}$). In all cases, $\mathfrak{g}_1 + L(C)/L(C)$ is a nonzero ideal in $L(G)$. By 14.26(i) it follows that $\mathfrak{g}_1 + L(C) = \mathfrak{g}_1 + K \cdot \mathrm{id} = L(G)$ if $p \neq 3$. If $p \neq 2, 3$ we even have $\mathfrak{g}_1 = L(G)$.

14.11 shows that $\mathfrak{h}_1$ is an ideal in $L(H)$, which is nonzero (one easily produces, by using the formulas describing $\mathscr{E}_3$, two elements $x, y \in V$ with $L(x,y) \neq 0$, $L(x,y)e = 0$). Hence 14.26(ii) shows $\mathfrak{h}_1 = L(H)$ if $p \neq 2$, in particular if $p = 3$. It now also follows that $\mathfrak{h}_1 = L(H)$ in all cases (use 14.10(i)). It also follows that $\mathfrak{g}_1 = L(G)$ if $p \neq 2$.

Finally we establish that $\mathfrak{g} = L(G)$. Let $D \in \mathfrak{g}$. By 14.9 we may assume that D induces a derivation of $L(G)$ which commutes with all $\mathrm{Ad}(t)$ ($t \in T$, a maximal torus of G). Application of 14.7(c) then shows that such a derivation is inner. Hence we may assume that D commutes with $L(G)$. But the action of the derived group G' in V is infinitesimally irreducible, i.e. $L(G')$ acts irreducibly in V (as follows from [2, 6.4, p. 42], applied to the 27-dimensional representation of G' in V). Hence D must be a multiple of the identity, whence $D \in L(G)$ and $\mathfrak{g} = L(G)$. By 14.10(ii) it follows that $L(H) = \mathfrak{h}$.

This concludes the case by case discussion of the various simple J-structures. We collect some of the results established in that discussion

§ 14. The Structure Group of a Simple J-structure

in the next theorems. We keep the previous notations. Recall that G is smooth if $\mathfrak{g}=L(G)$, similarly for H.

14.27 Theorem. *Let $\mathscr{S}$ be a simple J-structure.*
(i) *G is reductive and smooth;*
(ii) *H° is semisimple if and only if $\mathscr{S}$ is strongly simple and not of type $\mathscr{O}_{2,3}$;*
(iii) *H is smooth if and only if $\mathscr{S}$ is strongly simple.*

14.27(iii) is of importance for the treatment of rationality problems to be given in the next section.

14.28 Theorem. *Let $\mathscr{S}$ be a simple J-structure.*
(i) *Assume that $p \neq 2$ and that $\mathscr{S}$ is not of type $\mathscr{M}_r$, with p dividing r. Then $L(G) = \mathfrak{g}_1$. In the excepted cases $\mathfrak{g}_1$ has codimension 1 in $L(G)$;*
(ii) *If $\mathscr{S}$ is strongly simple and not of type $\mathscr{M}_r$ with $p \neq 2$ and p not dividing r, then $\mathfrak{h} = L(H) = \mathfrak{h}_1$. In particular, all derivations of $\mathscr{S}$ are then inner.*

14.29. Let $\mathscr{S} = (V, j, e)$ be a simple J-structure with inner structure group G_1. We know, by 14.13 and the remark following it, that G_1 is the identity component G° of the structure group G. By 14.27(i), G_1 is reductive.

Let A be the algebraic group constructed from $\mathscr{S}$ in 2.25 (there denoted by G). The assumptions (I) and (II) of 2.25 are satisfied here, hence A is a reductive group, which is in fact semisimple and adjoint. We shall identify A in the various cases.

First observe that from the construction of 2.25 it follows that A contains a parabolic subgroup Q isomorphic to the semidirect product $G_1 \cdot V$ (G_1 acting on V in the natural way). Let T be a maximal torus of G_1, let Σ be the root system of A with respect to T. Fix an ordering of Σ such that the corresponding Borel subgroup of A is contained in Q. Let Δ be the basis of Σ defined by the ordering.

Since the center C of G_1 is 1-dimensional (viz. the torus of nonzero scalar multiplications of V), it follows from the explicit description of standard parabolic subgroups given in [4, p. 85–86] that there is a root $\alpha \in \Delta$ such that Q is the standard parabolic subgroup $P_{\Delta - \{\alpha\}}$ defined by α. Q is a maximal parabolic subgroup of A.

14.30 Lemma. *A is a quasi-simple linear algebraic group.*

We have $G_1 = LC$, where L is either semisimple, or reduced to the neutral element (namely if $\mathscr{S} \simeq \mathscr{M}_1$). Assume first that L is quasi-simple. From the description of parabolic subgroups it then follows that either A is quasi-simple or A is isogeneous to a product $B \times \mathbb{SL}_2$, where B is quasi-simple.

In the second case L is isogeneous to B. Then $\dim A = \dim B + 3 = \dim L + 3 = \dim G_1 + 2$. But from the construction of 2.25 we see that $\dim A = \dim G_1 + 2 \dim V$. Hence we had $\dim V = 1$, $\mathcal{S} \simeq \mathcal{M}_1$, a contradiction.

If L is not quasi-simple, then either $\mathcal{S} \simeq \mathcal{M}_1$, in which case A is isomorphic to $\mathbb{PSL}_2$ (hence simple) or $\mathcal{S} \simeq \mathcal{M}_r$ with $r > 1$. Then L is isogeneous to $\mathbb{SL}_r \times \mathbb{SL}_r$. Again using the decription of parabolic groups, we see that A is isogeneous to a product $B \times \mathbb{SL}_r$, where B is either quasi-simple, or $B = \mathbb{SL}_2$. In this case, $\dim A = \dim G_1 + 2 \dim V = 4r^2 - 1$, and rank $A = 2r - 1$. Hence $\dim B = 3r^2$, rank $B = r$.

A check of the list of simple algebraic groups shows that for $r > 1$ a quasi-simple B with rank 2 and dimension r^2 does not exist (see [10, exposé 19]). If B is not quasi-simple, then $B = \mathbb{SL}_r \times \mathbb{SL}_r$ and $\dim B = r^2 + 2$, which cannot be equal to $3r^2$ for $r > 1$. Hence A is quasi-simple in all cases, which establishes 14.30.

14.31. We can now determine completely the structure of A. By 14.30 and by what we observed in 2.25, we know that A is an adjoint simple linear algebraic group. The structure of A is then completely determined if its type, i.e. that of the root system Σ, is given.

The type of A is listed below. We also have listed in the various cases, the root $\alpha \in \Delta$ which determines the parabolic group $Q = P_{\Delta - \{\alpha\}}$. The numbering of the roots of Δ is as in [7, p. 250-274].

Type of $\mathcal{S}$		Type of A	α
$\mathcal{M}_r$	$(r \geq 1)$	A_{2r}	α_2
$\mathcal{S}_r$	$(r \geq 2)$	C_r	α_r
$\mathcal{A}_r$	$(r \geq 3)$	D_{2r}	α_{2r-1} or α_{2r}
$\mathcal{E}_3$		E_7	α_7
$\mathcal{O}_{2,2r}$	$(\text{char}(K) \neq 2, r \geq 3)$	D_{r+1}	α_1
$\mathcal{O}'_{2,r}$	$(\text{char}(K) = 2, r \geq 3)$		
$\mathcal{O}_{2,2r+1}$	$(\text{char}(K) \neq 2, r \geq 2)$	B_{r+1}	α_1
$\mathcal{O}''_{2,r}$	$(\text{char}(K) = 2, r \geq 2)$		

The proof of the results contained in this list is easy. First one determines, for each particular $\mathcal{S}$, the dimension and the rank of A, which are given by $\dim A = \dim G_1 + 2 \dim V$, rank $A = $ rank G_1. G_1 ($=G^\circ$) is described above, in the various cases. Except when $\mathcal{S} = \mathcal{S}_r$, one checks, using the classification of simple algebraic groups, that $\dim A$ and rank A determine the type of A uniquely.

If $\mathcal{S} = \mathcal{S}_r$, the classification shows that A is either of type B_r or of type C_r. In that case G_1 is isogeneous to $\mathbb{GL}_r$. Let T be the torus in G_1 defines by the diagonal torus in $\mathbb{GL}_r$. The weights of T in V, considered as roots of Σ, can only occur in a root system of type C_r, so that Δ must be of type C_r.

The type of L determines the root $\alpha \in \Delta$ uniquely, except when $\mathscr{S} = \mathscr{A}_r$. Then we have the two possibilities shown above, related via the outer automorphism which interchanges α_{2r-1} and α_{2r}.

Notes

In the literature on Jordan algebras results like 14.27 and 14.28 are proved under restrictive assumption on char (K). For example, it is established in [8, IX, 3.7, p. 285], by a consideration of Killing forms, that in characteristic 0 the derivation algebra of a simple Jordan algebra A of dimension > 3 is semisimple, from which it follows that then the automorphism group of A is also semisimple.

For derivations of Jordan algebras see also [14, p. 253–258]. It is likely that 14.27(i) admits a converse, viz. that $\mathscr{S}$ is simple if G is reductive and smooth.

In the course of the case by case discussion given in §14 we have also determined the number of components of G and H. In characteristic 0 this was done in another way in [13].

§15. Rationality Questions

15.1. Let k be a subfield of K. We denote by k_s a separable closure of k, contained in K. Put $p = \text{char}(K)$.

$\mathscr{S} = (V, j, e)$ denotes a J-structure which is defined over k. Another such J-structure $\mathscr{S}' = (V', j', e')$ is called a *k-form* of $\mathscr{S}$ if it is K-isomorphic to $\mathscr{S}$.

Let H be the automorphism group of $\mathscr{S}$. If H is smooth in the sense of 4.9 it follows by 4.13 that a k-form $\mathscr{S}'$ of $\mathscr{S}$ is already k_s-isomorphic to $\mathscr{S}$. This permits the use of Galois cohomology. We recall some facts about this. For more details see [27].

15.2. Let V be a vector space which is defined over k. Let H be an algebraic subgroup of $GL(V)$, which is defined over k. We recall the definition of the 1-cohomology set $H^1(k, H)$.

Let Γ be the Galois group of k_s/k. This is a profinite group, which acts continuously on the group $H(k_s)$ of k_s-rational points of H (provided with the discrete topology). A 1-cocycle z of Γ in $H(k_s)$ is a continuous function $s \mapsto z_s$ of Γ into $H(k_s)$, such that

$$z_{st} = z_s \cdot s(z_t) \quad (s, t \in \Gamma).$$

Two cocycles z and z' are equivalent if there exists $h \in H(k_s)$ with

$$z'_s = h^{-1} z_s \cdot (s h).$$

The corresponding set of equivalence classes is denoted by $H^1(k, H)$. This set has a priviliged element 0, coming from the constant cocycle $z_s = 1$. We have $H^1(k, GL(V)) = 0$ (see [27, Lemma 1, p. III-3]).

The classification of k-forms of J-structures with smooth automorphism group is now given by the following result.

15.3 Proposition. *Let $\mathscr{S} = (V, j, e)$ be a J-structure which is defined over k, whose automorphism group H is smooth. Then H is defined over k. There is a bijection of the set of k-isomorphism classes of k-forms of $\mathscr{S}$ onto $H^1(k, H)$.*

§ 15. Rationality Questions

That H is defined over k was already established in 4.12. The last point is an example of a well-known principle, see [27, p. III-1]. We shall sketch the proof.

Let Σ be the set of isomorphism classes occurring in the assertion. Assume that $\mathscr{S}'=(V',j',e')$ is a k-form of $\mathscr{S}$. We may assume that $V'=V$. As we already observed in 15.1, there is an isomorphism g of $\mathscr{S}$ onto $\mathscr{S}'$ with $g \in GL(V)(k_s)$.

Let Γ be, as before, the Galois group of k_s/k. Put $z_s = g^{-1} \cdot (s\,g)$ $(s \in \Gamma)$. Then z is a 1-cocycle of Γ in $H(k_s)$, whose class depends only on the class of $\mathscr{S}'$ in Σ. This gives a map $\alpha: \Sigma \to H^1(k, H)$, which is readily seen to be injective.

If z is an arbitrary 1-cocycle of Γ in $H(k_s)$, there is, since $H^1(k, GL(V))=0$, an element $g \in GL(V)(k_s)$ with $z_s = g^{-1} \cdot (s\,g)$. We have $s(g\,e)=g\,e(s \in \Gamma)$, whence $g\,e \in V(k)$.

Put $e'=g\,e$ and
$$j' = g \circ j \circ g^{-1}.$$
It follows that j' is a birational map of V, which is defined over k and that $\mathscr{S}'=(V',j',e')$ is a k-form of $\mathscr{S}$ whose isomorphism class is mapped by α onto the cohomology class of z. This proves the bijectivity of α.

The aim of this section is to describe the k-forms of strongly simple J-structures. We first deal with the easy case of simple J-structures of degree 2.

15.4 Proposition. *Let $\mathscr{S}$ be a simple J-structure of degree 2 which is defined over k. There is a quadratic form Q on V, defined over k such that $Q(e)=1$ and such that $\mathscr{S}$ is k-isomorphic to the J-structure $\mathscr{J}(V,Q,e)$.*

The latter $\mathscr{J}$-structure was defined in §2. The proof of 15.4 is contained in 5.1, except for the assertions about k. These follow at once from what we established in 5.1.

15.5 Simple algebras with involution. We next recall some facts about simple algebras with involution. A simple algebra with involution is an algebra with involution (A, ρ) in the sense of 2.7 such that A has no nontrivial ρ-stable twosided ideals. It is known that one then has the following possibilities for (A, ρ), up to K-isomorphism:

(a) $A = \mathbb{M}_r \oplus \mathbb{M}_r$, $\rho(X, Y) = ({}^tY, {}^tX)$,
(b) $A = \mathbb{M}_{2r}$, $\rho X = S \cdot {}^tX \cdot S^{-1}$, where $S = \begin{pmatrix} 0 & 1_r \\ -1_r & 0 \end{pmatrix}$ (1_r = identity matrix),
(c) $A = \mathbb{M}_r$, $\rho X = {}^tX$.

This is proved in [33] (the restriction on the characteristic made there is unnecessary), see also [15, Ch. 0] for a more general case.

If (A, ρ) is a simple algebra with involution then the J-structure $\mathscr{J}(A,\rho)$ discussed in 2.8 is defined, and is isomorphic to $\mathscr{M}_r$, $\mathscr{S}_r$, $\mathscr{A}_r$, respectively, in the three cases just mentioned.

We shall say that a simple algebra with involution is *strongly simple* if either $p \neq 2$ or if $p = 2$ and we have one of the above cases (a), (b).

The automorphism group of (A, ρ) is in the three cases as follows:
(a) the group of linear transformations $(X, Y) \mapsto (TX \cdot T^{-1}, ({}^tT)^{-1} \cdot X \cdot {}^tT)$, with $T \in \mathbb{GL}_r$,
(b) the group of linear transformations $X \mapsto TXT^{-1}$, with T in the symplectic group $\mathbb{Sp}_{2r} = \{T \in \mathbb{GL}_r \mid TS \cdot {}^tT = S\}$,
(c) the group of linear transformations $X \mapsto TXT^{-1}$ with T in the orthogonal group $\mathbb{O}_r = \{T \in \mathbb{GL}_r \mid T \cdot {}^tT = 1\}$.

The next theorem now gives the k-forms of strongly simply J-structures of degree > 2, which are not exceptional, i.e. not isomorphic to $\mathscr{E}_3$.

15.6 Theorem. *Let $\mathscr{S}$ be a non-exceptional strongly simple J-structure of degree >2, which is defined over k. Then there exists a strongly simple algebra with involution (A, ρ), defined over k, such that $\mathscr{S}$ is k-isomorphic to $\mathscr{J}(A, \rho)$.*

$\mathscr{S}$ is isomorphic to either $\mathscr{M}_r$, $\mathscr{A}_r$ or $\mathscr{S}_r$ (the latter only if $p \neq 2$). Let (A', ρ') be a strongly simple algebra with involution over k, of one of the 3 types of 15.5, such that $\mathscr{S} \simeq \mathscr{J}(A', \rho')$. The results discussed in §14 and in 15.5 show that the automorphism group of $\mathscr{S}$ can be identified with that of (A', ρ'). 15.6 then follows, by an argument of the same kind as that used to prove 15.4. The details are left to the reader.

15.7. Let (A, ρ) be a simple associative algebra with involution, defined over k. We say that this is a simple algebra over k with an involution ρ *of the second kind* if (A, ρ) is in case (a) of 15.5 and if the algebra $A(k)$ of k-rational points is a simple algebra, whose center is a separable quadratic extension l of k. ρ then induces an involution of $A(k)$, which extends the nontrivial automorphism of l/k. We then have the following consequence of 15.6.

15.8 Corollary. *A k-form of $\mathscr{M}_r$ is either k-isomorphic to a J-structure $\mathscr{J}(A)$, where A is a simple associative algebra over k or to a J-structure $\mathscr{J}(A, \rho)$, where (A, ρ) is a simple algebra over k, with an involution ρ of the second kind.*

This follows from 15.6, using familiar facts about the k-classification of simple algebras with involution of case (a), see [33].

We finally discuss the exceptional J-structures. We shall prove the following result, due to Tits, about their k-forms. We use the notations of 5.13 and 5.16.

15.9 Theorem. *Let $\mathscr{S} = (V, j, e)$ be a k-form of the exceptional J-structure $\mathscr{E}_3$. Then $\mathscr{S}$ is k-isomorphic to an exceptional J-structure of the first kind or of the second kind, which is defined over k.*

§ 15. Rationality Questions

Let H be the automorphism group of $\mathscr{E}_3$. We know from 14.24 that H is a connected semisimple group of type F_4. Assume first that k is a finite field. Since H is connected, we have $H^1(k, H) = 0$, by a well-known theorem of Lang, see [27, p. III-14]. 15.3 then shows that $\mathscr{S}$ is k-isomorphic to $\mathscr{E}_3$, which establishes 15.9 in this case. Hence we can assume from now on that k is an *infinite* field.

We use the notations introduced in 5.2. In particular, N is the norm of $\mathscr{S}$ and n the quadratic map $V \to V$ which is the numerator of j. We then have the following formulas

(1) $\qquad n x \times n y + n(x \times y) = N(x, y) y + N(y, x) x,$

(2) $\qquad n x \times (x \times y) = N(x) y + N(x, y) x$

(3) $\qquad x \times (n x \times y) = N(x) y + \sigma(x, y) n x,$

where σ is the standard symmetric bilinear form of $\mathscr{S}$, introduced in 1.11. These are formulas (10), (11) and (12) of § 5.

Another formula we need is

(4) $\qquad x \times n x = -\sigma(e, x) n x - \sigma(e, n x) x + \bigl(\sigma(e, x) \sigma(e, n x) - N(x)\bigr) e.$

(4) follows by using 8.3, which shows that we have

$$x^3 = N(e, x) x^2 - N(x, e) x + N(x) e,$$

and using § 5, (7) (with $y = e$ and $y = x$).

Let $x, y \in V$. There is a smallest substructure $\mathscr{S}_{x,y}$ of $\mathscr{S}$ containing e, x and y. From (1), (2) and (3) one sees that $\mathscr{S}_{x,y}$ is spanned by the elements e, $x, y, n(x), n(y), x \times n y, n x \times y, x \times y, n x \times n y$. Hence $\dim \mathscr{S}_{x,y} \le 9$. Moreover, we have $\dim \mathscr{S}_{x,y} = 9$ for suitable x, y.

In fact, $\mathscr{S}$ is isomorphic to $\mathscr{E}_3$, which contains $\mathscr{M}_3$ as a substructure and it is easily seen that $\mathscr{M}_3$ can be generated by 2 elements. Also, for a suitable choice of x and y we have that $\mathscr{S}_{x,y}$ is a J-structure of degree 3, with a nondegenerate irreducible norm (viz. the J-structure $\mathscr{M}_3$).

It follows that there exists a nonempty open subset U of $V \times V$ such that for $(x, y) \in U$ the substructure $\mathscr{S}_{x,y}$ is a J-structure of degree 3 and dimension 9, with a nondegenerate irreducible norm. By 13.8 it then follows that $\mathscr{S}_{x,y}$ is simple, and it can then only be isomorphic to $\mathscr{M}_3$ (use 13.6). Since k is infinite we may choose x and y to be in $V(k)$. Then $\mathscr{S}_{x,y}$ is a k-form of $\mathscr{M}_3$. These are described in 15.8.

First assume that $\mathscr{S}_{x,y}$ is isomorphic to $\mathscr{J}(A)$, where A is a simple associative algebra, defined over k. We then identify $\mathscr{S}_{x,y}$ with A, so A is now a k-subspace of V and $j a = a^{-1}$ if a is an invertible element of A. The restriction of N to A must be the norm of $\mathscr{J}(A)$. From 2.4 we then see that the restriction of σ to $A \times A$ is nondegenerate.

Let W be the orthogonal complement of A with respect to σ, i.e.

$$W = \{x \in V \mid \sigma(x, A) = 0\}.$$

If $a \in A$, define a linear transformation $\phi(a)$ of W into V by

$$\phi(a) x = -a \times x \quad (x \in W).$$

From §5, (5) it follows that this defines a linear transformation of W into itself. Moreover §5, (7) implies that $\phi(e) = \mathrm{id}$. From (3) it follows that if $a \in A$ is invertible, $\phi(a)$ is an invertible linear transformation of W and that then

$$\phi(a)^{-1} = \phi(a^{-1}).$$

We conclude from 2.6 that there is a direct sum decomposition $W = V_1 \oplus V_2$, with

$$V_1 = \{x \in W \mid \phi(ab) x = \phi(a) \phi(b) x \text{ if } a, b \in A\},$$
$$V_2 = \{x \in W \mid \phi(ba) x = \phi(a) \phi(b) x \text{ if } a, b \in A\},$$

moreover V_1 and V_2 are defined over k.

Next we prove that $\sigma(V_i, V_i) = 0$ $(i = 1, 2)$. Let $a, b \in A$, $x, y \in V_1$. Using §5, (5) we find that

$$\sigma((ab) \times x, y) = -\sigma(a \times (b \times x), y) = -\sigma(x, b \times (a \times y))$$
$$= \sigma(x, (ba) \times y) = \sigma((ba) \times x, y).$$

Hence $\sigma((ab - ba) \times x, y) = 0$ and it follows that

$$\sigma(\phi(a) x, y) = 0$$

for all $a \in A$ with trace 0. For an invertible a with this property we have that the map $x \mapsto \phi(a) x$ is bijective. This implies that $\sigma(V_1, V_1) = 0$, and similarly $\sigma(V_2, V_2) = 0$.

Since σ is nondegenerate it follows from $\sigma(V_1, A) = 0$, $\sigma(V_1, V_1) = 0$ that for nonzero $y \in V_2$ the linear function $x \mapsto \sigma(x, y)$ on V_1 is nonzero. Hence $\dim V_2 \leq \dim V_1$, and also $\dim V_2 \leq \dim V_1$, so that $\dim V_1 = \dim V_2 = 9$.

The simple associative algebra A acts in V_1 by $(a, x) \mapsto \phi(a) x$ and it is well-known that this action is completely reducible (over K). Hence V_1 can be decomposed into the direct sum of 3 A-stable subspaces of dimension 3, in each of which the action is isomorphic to that of $\mathbb{M}_3$ in K^3. It then follows that there exists $v_1 \in V_1$ such that the map $a \mapsto \phi(a) v_1$ of A into V_1 is *injective*.

Let $x, y \in V_i$, $a \in A$. Then $\sigma(x \times y, a) = \sigma(x, a \times y) = 0$. Hence $V_i \times V_i \subset W$. If $p \neq 2$ we have $nx = \frac{1}{2} x \times x \in W$, whence $nV_i \subset W$. We shall prove this

§ 15. Rationality Questions

also for $p=2$, more precisely we shall show that (in all cases) $nV_1 \subset V_2$, $nV_2 \subset V_1$.

From (4) we obtain, if $x \in V_1$, $a \in A$

(5) $$a \times nx + x \times (x \times a) = -\sigma(e,a)nx - \sigma(e,nx)a \\ + (\sigma(e,a)\sigma(e,nx) - \sigma(nx,a))e,$$

whence using §5, (5) and $\sigma(V_1, V_1) = 0$,

(6) $$\sigma(a \times b, nx) = -\sigma(e,a)\sigma(b,nx) - \sigma(e,nx)\sigma(a,b) \\ + \sigma(e,a)\sigma(e,b)\sigma(e,nx) - \sigma(e,b)\sigma(a,nx),$$

if $x \in V_1$, $a,b \in A$.

Now from 8.2(i) we obtain that

$$na = a^2 - \sigma(e,a)a + \sigma(na,e)e \quad (a \in A),$$

and

(7) $$a \times b = ab + ba - \sigma(e,a)b - \sigma(e,b)a + (\sigma(e,a)\sigma(e,b) - \sigma(a,b))e$$

(using $e \times a = -a + \sigma(e,a)e$). Putting this into (6) we see that

(8) $$\sigma(nx, ab+ba) = 0 \quad (x \in V_1, a,b \in A).$$

Let c be the component of nx in A. It then follows readily from (8) that c must be a multiple of e. Hence there is a quadratic form Q on V_1 such that $nx - Q(x)e \in W$.

From (3) we obtain, if $a,b \in A$, $x \in V_1$ that

$$a \times (b \times nx) + x \times (b \times (a \times x)) = \sigma(a, nx)b + \sigma(a,b)nx.$$

Now

$$x \times (b \times (a \times x)) = -x \times ((ba) \times x),$$

using (5) we then obtain

$$a \times (b \times nx) = -ba \times nx + \sigma(nx, a)b - \sigma(e, nx)ba \\ + (\sigma(e, ba)\sigma(e, nx) - \sigma(nx, ba))e.$$

This implies that $nx - Q(x)e \in V_2$. Moreover, if $Q(x) \neq 0$ for some $x \in V_1$, it would follow that

$$a \times (b \times e) = -ba \times e + \sigma(e,a)b - \sigma(e,e)ba + 2\sigma(e,ba)e.$$

Using that we must have $p=2$ it then follows from (8) that $ab = ba$ for all $a, b \in A$, a contradiction. Hence $Q=0$ and $nV_1 \subset V_2$. Similarly $nV_2 \subset V_1$. Since $\sigma(V_1 \times V_2, V_1) = \sigma(V_2, V_1 \times V_1) = 0$, we have $V_1 \times V_2 \subset A$.

Now let $v_1 \in V_1$ be such that $a \mapsto \phi(a)v_1$ is injective map of A onto V_1. Such v_1 exist, and since k is infinite we may and shall assume that $v_1 \in V_1(k)$.

If $a \in A$ is then such that $\sigma(a, v_1 \times V_2) = 0$, we have $\sigma(a \times v_1, V_2) = 0$, whence $\phi(a) v_1 = -a \times v_1 = 0$. Hence $a = 0$. It follows that $v_1 \times V_2 = A$.

Now we have for all $x \in V$ that

$$N(n\,x) = N(x)^2,$$

as was established in the beginning of the proof of 5.5. This implies, by a straightforward computation which we omit, that

$$N(x \times y) = -N(x) N(y) + \sigma(n\,x, y) \sigma(x, n\,y).$$

Take v_1 as before in V_1, and let $x \in V_2$. Then, since $n V_1 \subset V_2$, $n V_2 \subset V_1$, $\sigma(V_i, V_i) = 0$, we find

$$N(v_1 \times x) = -N(v_1) N(x).$$

Since $v_1 \times V_2 = A$, we may take x such that $N(v_1 \times x) \neq 0$. It follows that $N(v_1) \neq 0$. Put $\alpha = N(v_1)$ and let $v_2 = \alpha^{-1} n v_1$. Then $v_2 \in V_2(k)$ and $N(v_2) \neq 0$. Moreover, by (4),

$$v_1 \times v_2 = \alpha^{-1} v_1 \times n(v_1) = -e.$$

Since by (3) we have

$$v_1 \times (n v_1 \times a) = N(v_1) a,$$

it follows that $V_2 = \phi(A) v_2$. We have

$$N(\phi(a) v_1) = \alpha N(a), \quad N(\phi(a) v_2) = \alpha^{-1} N(a).$$

It follows that we have, if $x_0, x_1, x_2 \in A$,

$$N(x_0 + \phi(x_1) v_1 + \phi(x_2) v_2) = N(x_0) + \alpha N(x_1) + \alpha^{-1} N(x_2)$$
$$+ \sigma(\phi(x_1) v_1 \times \phi(x_2) v_2, x_0).$$

The last term equals, using §5, (5) and the definitions of V_1 and V_2,

$$\sigma((x_1 \times v_1) \times (x_2 \times v_2), x_0) = \sigma(x_1 \times v_1, x_0 \times (x_2 \times v_2))$$
$$= -\sigma(x_1 \times v_2, (x_2 x_0) \times v_2) = \sigma(v_1, (x_1 x_2 x_0) \times v_2)$$
$$= \sigma(v_1 \times v_2, x_1 x_2 x_0) = -\sigma(e, x_1 x_2 x_0) = -\tau(x_0 x_1 x_2),$$

where τ is the trace in A.

Consequently,

(9) $\quad N(x_0 + \phi(x_1) v_1 + \phi(x_2) v_2) = N(x_0) + \alpha N(x_1) + \alpha^{-1} N(x_2) - \tau(x_0 x_1 x_2),$

which shows that N is as in §5, (14). Since j is completely determined by N and e (see 12.3(iv)), it follows that $\mathscr{S}$ is k-isomorphic to the exceptional J-structure of the first kind $\mathscr{E}_3(A, \alpha)$ defined in 5.13.

We have now dealt with the case that the substructure $\mathscr{S}_{x,y}$ of the beginning of the proof of 15.8 is k-isomorphic to some $\mathscr{J}(A)$, A a simple asso-

§ 15. Rationality Questions

ciative algebra over k. There remains the case that $\mathscr{S}_{x,y}$ is k-isomorphic to a J-structure $\mathscr{J}(B,\rho)$, where B is a simple algebra over k with an involution ρ of the second kind.

Then there is a separable quadratic extension l of k, contained in K, and a simple associative algebra A defined over l and K-isomorphic to $\mathbb{M}_3$, together with an involution ι of A which is defined over l, such that $B = A \oplus A$ and that $\rho(a_1, a_2) = (\iota a_2, \iota a_1)$. l is the center of $B(k)$ and ρ extends the nontrivial automorphism $x \mapsto x'$ of l/k. We have that $\mathscr{S}_{x,y}$ is l-isomorphic to $\mathscr{J}(A)$, hence over l we are in the situation just discussed.

Consider $V(l)$. We may then identify $V(l)$ with $A(l)^3$, the norm N being given by (9), where α is a suitable element of l^*. The automorphism $x \mapsto x'$ of l/k defines a similar semilinear automorphism of $V(l)$, to be denoted in the same way. Remark that $(x_0, 0, 0)' = (\rho\, x_0, 0, 0)$ $\bigl(x_0 \in A(l)\bigr)$.

We claim that with V_1 and V_2 as before, we have $V_1(l)' = V_2(l)$, $V_2(l)' = V_1(l)$. Let for example $x \in V_1(l)$, $a, b \in A(l)$. Then

$$a \times (b \times x') = \bigl(\rho(a) \times (\rho(b) \times x)\bigr)' = -\bigl(\rho(b\,a) \times x\bigr)' = -(b\,a) \times x',$$

whence $V_1(l)' = V_2(l)$.

Put $(0, 1, 0)' = (0, 0, u)$. Since $N(x)' = N(x')$, we find that $N(u) = \alpha\,\alpha'$. Also, we have $(0, 0, 1)' = (0, v, 0)$ and using $(0, 0, 1) \times (0, 1, 0) = -(1, 0, 0)$ one finds that $v = u^{-1}$. Since $(0, 1, 0) = (0, 0, u')' = (0, u'\,u^{-1}, 0)$, we have $u' = u$.

It then follows that, if $x_i \in A(l)$, we have

$$(x_0, x_1, x_2)' = \bigl(\rho\, x_0, \rho(x_2)\, u^{-1}, u\,\rho(x_1)\bigr).$$

Since

$$V(k) = \{x \in V(l) \mid x' = x\}$$

we have that $V(k)$ can be identified with $B_0(k) \oplus B(l)$, where $B_0 = \{x \in B \mid \rho\, x = x\}$. Moreover, the norm on $V(k)$ is then given by

$$N(x_0 + x) = N(x_0) + \alpha N(x) + \alpha'\, N(x') - \tau(x_0 \, x \, u \, x').$$

This shows that on $V(k)$ the norm is given by a formula of the form of §5, (21). Since k is infinite, this determines N on V. We conclude, using 12.3 (iv), that $\mathscr{S}$ is isomorphic to the J-structure of the second kind $\mathscr{E}_3\,(B, \rho, u, \alpha)$ of 5.16. This finishes the proof of 15.9.

Notes

For the classification of special (= non-exceptional) Jordan algebras over a field of characteristic not 2, due to Jacobson, see [14, Ch. V, Theorem 11, p. 210]. For the characteristic 2 case see [14, p. 3.59].

Tits' proof of 15.9, which is unpublished, is more group theoretical than the one given here, which is an adaptation of a proof of McCrimmon, given in [24].

Bibliography

1. Borel, A.: Linear algebraic groups. New York: Benjamin 1969.
2. Borel, A. et al.: Seminar on algebraic groups and related finite groups. Lecture Notes in Mathematics, No. 131. Berlin-Heidelberg-New York: Springer 1970.
3. Borel, A., Springer, T. A.: Rationality properties of linear algebraic groups II. Tôhoku Math. J. (2), **20**, 443–497 (1968).
4. Borel, A., Tits, J.: Groupes réductifs. Publ. math. I.H.E.S. **27**, 55–150 (1965).
4a. Borel, A., Tits, J.: Homomorphismes «abstraits» de groupes algébriques simples. To appear in Ann. of Math.
5. Bourbaki, N.: Algèbre, 2nd ed., Chap. 4, 5. Paris: Hermann 1959.
6. Bourbaki, N.: Algèbre, Chap. 8. Paris: Hermann 1958.
7. Bourbaki, N.: Groupes et algèbres de Lie, Chap. 4, 5, 6. Paris: Hermann 1968.
8. Braun, H., Koecher, M.: Jordan-Algebren. Berlin-Heidelberg-New York: Springer 1966.
9. Chevalley, C.: Sur certains groupes simples. Tôhoku Math. J. (2), **7**, 14–66 (1955).
10. Chevalley, C.: Séminaire sur la classification des groupes de Lie algébriques, 2 Vol. Paris 1958.
11. Demazure, M., Grothendieck, A.: Schémas en groupes. Lecture Notes in Mathematics, Nos. 151, 152, 153. Berlin-Heidelberg-New York: Springer 1970.
12. Dieudonné, J.: La géometrie des groupes classiques, 2nd ed., Ergebnisse der Mathematik und ihrer Grenzgebiete, Bd. 5. Berlin-Göttingen-Heidelberg: Springer 1963.
13. Gordon, S. R.: The components of the automorphism group of a Jordan algebra. Trans. Amer. Math. Soc. **153**, 1–50 (1971).
14. Jacobson, N.: Structure and representations of Jordan algebras. American Mathematical Society Colloquium Publications, Vol. XXXIX. Providence, R. I.: American Mathematical Society 1968.
15. Jacobson, N.: Lectures on quadratic Jordan algebras. Tata Institute of Fundamental Research, Bombay, 1969.
16. Koecher, M.: Analysis in reellen Jordan-Algebren. Nachr. Akad. Wiss. Göttingen **1958**, 67–74.
17. Koecher, M.: On homogeneous algebras. Bull. Amer. Math. Soc. **72**, 347–357 (1966).
18. Koecher, M.: Über eine Gruppe von rationalen Abbildungen. Invent. Math. **3**, 136–171 (1967).
19. Koecher, M.: Imbedding of Jordan algebras into Lie algebras, I, II. Amer. J. Math. **89**, 787–816 (1967); **90**, 476–510 (1968).
20. Kostant, B.: A characterization of the classical groups. Duke Math. J. **25**, 107–123 (1958).
21. McCrimmon, K.: A general theory of Jordan rings. Proc. Nat. Acad. Sci. U.S.A. **56**, 1072–1079 (1966).
22. McCrimmon, K.: The radical of a Jordan algebra. Proc. Nat. Acad. Sci. U.S.A. **62**, 671–678 (1969).

23. McCrimmon, K.: The Freudenthal-Springer-Tits constructions of exceptional Jordan algebras. Trans. Amer. Math. Soc. **139**, 495–510 (1969).
24. McCrimmon, K.: The Freudenthal-Springer-Tits constructions revisited. Trans. Amer. Math. Soc. **148**, 293–314 (1970).
25. Rosenlicht, M.: Some basic theorems on algebraic groups. Amer. J. Math. **78**, 401–443 (1956).
26. Seligman, G.B.: Modular Lie algebras. Ergebnisse der Mathematik und ihrer Grenzgebiete, Bd. 40. Berlin-Heidelberg-New York: Springer 1967.
27. Serre, J.-P.: Cohomologie Galoisienne. Lecture Notes in Mathematics, No. 5. Berlin-Heidelberg-New York: Springer 1964.
28. Serre, J.-P.: Sur les groupes de Galois attachés aux groupes p-divisibles. In: Proceedings of a Conference on Local Fields, pp. 118–131. Berlin-Heidelberg-New York: Springer 1967.
29. Steinberg, R.: Automorphisms of classical Lie algebras. Pacific J. Math. **11**, 1119–1129 (1961).
30. Steinberg, R.: Endomorphisms of linear algebraic groups. Mem. Amer. Math. Soc. No. 80, 108 p. (1968).
31. Tits, J.: Une classe d'algèbres de Lie en relation avec les algèbres de Jordan. Proc. Kon. Akad. Wet. Amst. **65**, 530–535 (1962).
32. Weil, A.: On algebraic groups of transformations. Amer. J. Math. **77**, 355–391 (1955).
33. Weil, A.: Algebras with involution and the classical groups. J. Indian Math. Soc. **24**, 589–623 (1960).

Index

$\mathscr{A}_r$ 29
admissible triple 108

central idempotent 95
characteristic ideal 83
classification of simple J-structures 134
— of strongly simple J-structures 125

degree of a J-structure 11
derivation of a J-structure 51
derivative 6
differential 6
direct sum of J-structures 20

$\mathscr{E}_3$ 63
exceptional J-structure of the first kind 63
— of the second kind 64

ideal of a J-structure 11
idempotent element of a J-structure 45
inner derivation of a J-structure 51
— structure algebra 50
— structure group 18
invariant bilinear form 19
involution of the second kind 160

Jordan algebra 67
Jordan decomposition 82
J-structure 10
— defined by associative algebra 24
— by associative algebra with involution 28
— by Jordan algebra 70
— by quadratic form 32
— without identity 21

k-form of a J-structure 158

$\mathscr{M}_r$ 25

minimum polynomial 79
morphism of J-structures 10

nil ideal 87
nilpotent element of a J-structure 46
norm of a J-structure 11

$\mathcal{O}_{2,r}, \mathcal{O}'_{2,r}, \mathcal{O}''_{2,r}$ 33
orthogonal idempotents 97

Peirce decomposition 92
polynomial function 2
— map 2
powers of an element in a J-structure 45
primitive idempotent 97

quadratic form 8
— Jordan algebra 72
— map of a J-structure 42
quotient of J-structure by an ideal 84

radical 86
— element 86
rational function 3
— map 3

$\mathscr{S}_r$ 28
semi-invariant 15
semisimple element of a J-structure 81
— J-structure 86
separable J-structure 74
— quadratic Jordan algebra 74
simple algebra with involution 159
— J-structure 11
smooth 52
standard automorphism 11
— symmetric bilinear form 14
strongly simple J-structure 122
structure algebra 49
structure group of a J-structure 10
substructure generated by an element 81

Ergebnisse der Mathematik und ihrer Grenzgebiete

1. Bachmann: Transfinite Zahlen. DM 48,—; US $15.30
2. Miranda: Partial Differential Equations of Elliptic Type. DM 58,—; US $18.40
4. Samuel: Méthodes d'Algèbre Abstraite en Géométrie Algébrique. DM 34,—; US $10.80
5. Dieudonné: La Géométrie des Groupes Classiques. DM 42,—; US $13.40
7. Ostmann: Additive Zahlentheorie. 1. Teil: Allgemeine Untersuchungen. DM 42,—; US $13.40
8. Wittich: Neuere Untersuchungen über eindeutige analytische Funktionen. DM 36,—; US $13.40
11. Ostmann: Additive Zahlentheorie. 2. Teil: Spezielle Zahlenmengen. DM 34,—; US $10.80
13. Segre: Some Properties of Differentiable Varieties and Transformations. DM 46,—; US $14.60
14. Coxeter/Moser: Generators and Relations for Discrete Groups. DM 42,—; US $13.40
15. Zeller/Beckmann: Theorie der Limitierungsverfahren. DM 64,—; US $20.30
16. Cesari: Asymptotic Behavior and Stability Problems in Ordinary Differential Equations. DM 54,—; US $17.20
17. Severi: Il theorema di Riemann-Roch per curve – superficie e varietà questioni collegate. DM 30,—; US $9.60
18. Jenkins: Univalent Functions and Conformal Mapping. DM 37,—; US $11.80
19. Boas/Buck: Polynomial Expansions of Analytic Functions. DM 24,—; US $7.70
20. Bruck: A Survey of Binary Systems. DM 46,—; US $14.60
21. Day: Normed Linear Spaces. In preparation
23. Bergmann: Integral Operators in the Theory of Linear Partial Differential Equations. DM 40,—; US $12.70
25. Sikorski: Boolean Algebras. DM 42,—; US $13.40
26. Künzi: Quasikonforme Abbildungen. DM 43,—; US $13.70
27. Schatten: Norm Ideals of Completely Continuous Operators. DM 30,—; US $9.60
28. Noshiro: Cluster Sets. DM 40,—; US $12.70
30. Beckenbach/Bellman: Inequalities. DM 38,—; US $12.10
31. Wolfowitz: Coding Theorems of Information Theory. DM 30,—; US $9.60
32. Constantinescu/Cornea: Ideale Ränder Riemannscher Flächen. DM 75,—; US $23.80
33. Conner/Floyd: Differentiable Periodic Maps. DM 34,—; US $10.80
34. Mumford: Geometric Invariant Theory. DM 24,—; US $7.70
35. Gabriel/Zisman: Calculus of Fractions and Homotopy Theory. DM 42,—; US $13.40
36. Putnam: Commutation Properties of Hilbert Space Operators and Related Topics. DM 31,—; US $9.90
37. Neumann: Varieties of Groups. DM 51,—; US $16.20
38. Boas: Integrability Theorems for Trigonometric Transforms. DM 20,—; US $6.20
39. Sz.-Nagy: Spektraldarstellung linearer Transformationen des Hilbertschen Raumes. DM 24,—; US $7.70
40. Seligman: Modular Lie Algebras. DM 43,—; US $13.70
41. Deuring: Algebren. DM 30,—; US $9.60
42. Schütte: Vollständige Systeme modaler und intuitionistischer Logik. DM 30,—; US $9.60
43. Smullyan: First-Order Logic. DM 36,—; US $11.50
44. Dembowski: Finite Geometries. DM 68,—; US $21.60
45. Linnik: Ergodic Properties of Algebraic Fields. DM 44,—; US $14.00
46. Krull: Idealtheorie. DM 34,—; US $10.80
47. Nachbin: Topology on Spaces of Holomorphic Mappings. DM 18,—; US $5.80

48. A. Ionescu Tulcea/C. Ionescu Tulcea: Topics in the Theory of Lifting. DM 36,—; US $11.50
49. Hayes/Pauc: Derivation and Martingales. DM 48,—; US $15.30
50. Kahane: Séries de Fourier absolument convergentes. DM 44,—; US $14.00
51. Behnke/Thullen: Theorie der Funktionen mehrerer komplexer Veränderlichen. DM 48,—; US $15.30
52. Wilf: Finite Sections of Some Classical Inequalities. DM 28,—; US $8.90
53. Ramis: Sous-ensembles analytiques d'une variété banachique complexe. DM 36,—; US $11.50
54. Busemann: Recent Synthetic Differential Geometry. DM 32,—; US $10.20
55. Walter: Differential and Integral Inequalities. DM 74,—; US $23.50
56. Monna: Analyse non-archimédienne. DM 38,—; US $12.10
57. Alfsen: Compact Convex Sets and Boundary Integrals. DM 46,—; US $14.60
58. Greco/Salmon: Topics in m-Adic Topologies. DM 24,—; US $7.70
59. López de Medrano: Involutions on Manifolds. DM 36,—; US $11.50
60. Sakai: C*-Algebras and W*-Algebras. DM 68,—; US $21.60
61. Zariski: Algebraic Surfaces. DM 54,—; US $17.20
62. Robinson: Finiteness Conditions and Generalized Soluble Groups, Part 1. DM 48,—; US $15.30
63. Robinson: Finiteness Conditions and Generalized Soluble Groups, Part 2. DM 64,—; US $20.30
64. Hakim: Topos annelés et schémas relatifs. DM 48,—; US $15.30
65. Browder: Surgery on Simply-Connected Manifolds. DM 42,—; US $13.40
66. Pietsch: Nuclear Locally Convex Spaces. DM 48,—; US $15.30
67. Dellacherie: Capacités et processus stochastiques. DM 44,—; US $14.00
68. Raghunathan: Discrete Subgroups of Lie Groups. DM 56,—; US $17.80
69. Rourke/Sanderson: Introduction to Piecewise-Linear Topology. DM 42,—; US $13.40
70. Kobayashi: Transformation Groups in Differential Geometry. DM 52,—; US $16.50
71. Tougeron: Idéaux de fonctions différentiables. DM 69,—; US $21.90
72. Gihman/Skorohod: Stochastic Differential Equations. DM 88,—; US $27.90
73. Milnor/Husemoller: Symmetric Bilinear Forms. DM 42,—; US $13.40
74. Fossum: The Divisor Class Group of a Krull Domain. DM 44,—; US $14.00
75. Springer: Jordan Algebras and Algebraic Groups. DM 48,—; US $15.30
76. Wehrfritz: Infinite Linear Groups. DM 59,—; US $18.70

Prices are subject to change without notice

Springer and the environment

At Springer we firmly believe that an international science publisher has a special obligation to the environment, and our corporate policies consistently reflect this conviction.

We also expect our business partners – paper mills, printers, packaging manufacturers, etc. – to commit themselves to using materials and production processes that do not harm the environment. The paper in this book is made from low- or no-chlorine pulp and is acid free, in conformance with international standards for paper permanency.

Printing and binding: Druckerei Triltsch, Würzburg